ON THE ROCKS

A Geology of Britain

Written and edited by Dr Robert Muir Wood
of the Department of Mineralogy and Petrology,
University of Cambridge

Gazetteer section of Sites of Geological interest
in England, Wales and Scotland compiled by
Peter R Rodgers

D0269966

BRITISH BROADCASTING CORPORATION

This book accompanies the BBC
Further Education Television programmes
On the Rocks, first broadcast on BBC2
from 21 February 1978

The series is produced by Brenda Horsfield

Published to accompany a series of
programmes prepared in consultation with the
BBC Further Education Advisory Council

This book is set in Monophoto Bembo 11/12pt

First published 1978. Reprinted 1978
Published by the British Broadcasting
Corporation
35 Marylebone High Street, London
W1M 4AA

Printed by Ebenezer Baylis & Son Ltd.,
Leicester and London.

ISBN 0 563 16211 2

Contents

Acknowledgment is due to the following for permission to reproduce photographs:

AEROFILMS LTD plates 2.7, 3.8, 9.4, 9.5, 9.6, 9.12, G.4, G.15; BARNABY'S PICTURE LIBRARY front cover and plates 1.19, 3.13, 7.1, 7.3, 7.17, 7.21, 7.24, 8.12, 9.2, 9.15, 10.1, 10.2, 10.3, 10.4, 10.6, 10.9, 10.10, 10.11, G.9, G.18; BRITISH MUSEUM plate 10.16; BRITISH PETROLEUM CO. LTD plates 8.9, 8.13; BRITISH TOURIST AUTHORITY colour plate 21; BUNDESBILDSTELLE, BONN plate 8.3; CAMBRIDGE UNIVERSITY COLLECTION plates 2.9, 9.10; CAMERA PRESS LTD plates 1.16, 7.6, 10.7, 10.13, G.2, G.3, G.5, G.8, G.13; BRUCE COLEMAN LTD plate 8.2; CAPT. E. D. J. EVANS plate 3.17; FOTO MAS plate 5.5; J. GRIFFITHS plate 2.2; J. HANCOCK plate 7.16; HAWAII NATURAL HISTORY ASSOCIATION colour plate 1; NICHOLAS HORNE LTD plate 4.10; BRENDA HORSFIELD plate 2.16; I.C.I. plate 7.19; THE DIRECTOR OF THE INSTITUTE OF GEOLOGICAL SCIENCES (N.E.R.C. COPYRIGHT) plates 1.10, 1.11, 1.12, 1.13, 1.18, 1.20, 2.1, 2.3, 2.4, 2.5, 2.11, 2.13, 2.14, 2.15, 2.17, 2.21, 2.25, 3.7, 4.1, 4.3, 4.4, 4.8, 4.9a, 4.9b, 4.11, 4.14, 4.15, 5.8, 5.9, 5.11, 5.12, 5.13, 5.14, 5.15, 6.5, 6.6, 6.8, 6.9, 6.11b, 6.11c, 6.11d, 6.11e, 6.13b, 6.13c, 6.13d, 6.13e, 6.13f, 6.16, 6.18, 7.7, 7.10, 7.11, 7.12, 7.13, 7.14, 7.22, 7.23, 8.4d, 8.4e, 9.17, 9.18, 9.19, 9.20, 9.21, 9.22, 9.23, 9.24, 9.25, 9.28, colour plates 3, 5, 6, 7, 8, 9, 10, 11, 12, 13, 14 & 18; IRISH TOURIST BOARD plates 8.1, 9.16; LEICESTER CITY MUSEUM plate 6.7; D.C. THOMPSON AND CO. LTD plate G.17; NORWEGIAN TRAVEL BUREAU plate 9.1; POPPERFOTO plate 10.12; PACE plates 6.11a, 8.4a, 8.4b, 8.4c; PERMUTIT CO. LTD plate 7.4; PILKINGTON BROTHERS LTD plate 5.1; PHOTO LIBRARY OF AUSTRALIA plate 10.8; RADIO TIMES HULTON PICTURE LIBRARY plate 3.5; PETER R. RODGERS plates G.1, G.7, G.10, G.11, G.12, G.16, G.19 and colour plates 15 & 16; SCOTTISH DEVELOPMENT DEPARTMENT (CROWN COPYRIGHT RESERVED) plate 1.7; SCOTTISH TOURIST BOARD plate G.20 and colour plate 4; SPACE FRONTIERS LTD (NASA) plate 2.10; STEWARTS & LLOYD LTD plate 7.20; STUDIO JON plate G.6; SWISS NATIONAL TOURIST OFFICE plate 9.7; JOHN TOPHAM PICTURE LIBRARY plates 4.6, 10.14, G.21; TRANS ANTARCTIC EXPEDITION plate 9.13; UNITED STATES GEOLOGICAL SURVEY plates 1.6, 3.3, 4.7, 9.8; VICTORIA & ALBERT MUSEUM (CROWN COPYRIGHT) plate 1.1; BRADFORD WASHBURN plate 9.3; JAMES WEIR colour plate 20; C.H. WOOD (BRADFORD) LTD plate G.14.

The drawings are by John Gilkes, Brian and Constance Dear, and Hugh Ribbans.

Introduction

For over 200 years Geology has been a subject of serious study in Britain. So much pioneering work was done here that the names of some Geological formations now in use all over the world, have come from an English county (Devon) and two ancient Welsh tribes (the Silures and the Ordovices).

Sadly all this scholarly effort has not been available to most of the people who live in these islands. It has been inherited by a privileged and mainly professional few, since Geology is taught at a basic level in only a handful of schools, and as a subject of higher education is best approached by students who already possess a fairly advanced knowledge of chemistry, physics and mathematics. In consequence a lot of people are unnecessarily deprived of information that would add enormously to their enjoyment of their environment – the places in which they live, work, travel and take their holidays.

There is another reason for the sparseness of geological knowledge. The Geology of Britain is varied and complex; the country is a cluster of 400 islands, many of which have a geological history that is still not fully investigated or understood. So the rocks closest to home may not be the most rewarding material for the beginner. In the last decade, however, a new wave of simplification has swept over Geology with the digestion of data exchanged with the sister sciences of Geophysics and Oceanography. Suddenly a lot of knotty British puzzles are being unravelled with the help of such studies as *plate tectonics* and *palaeo-magnetism* coupled with improved techniques of rock dating.

Now a great deal of classical geological detail can be fitted into a new broad framework. What seemed, not long ago, like well-trodden, even sterile ground has, once again, the feel of the frontier. It is in the belief that almost everybody can share in the excitement of the new work and new ideas if they are made accessible, that this book has been written and the BBC Television series of the same name, (for which it was designed as a partner), was made.

One of the best ways to start developing 'geological eyes' in the countryside is to go there with a trained geologist who can identify and explain the special features of the landscape and how they got there.

Each of the TV programmes in the series was built round such a field experience, but memory fades and the expert guide 'on the box' is not available to answer the viewer's own questions. Part of the function of this book is to supply some of the information that is bound to be needed by anyone attempting to go over the same ground alone. In fact it offers a good deal more material than can be contained in half-hour programmes. Detailed explanations and the reasons behind some current interpretations have been presented in 'colour panels' for more advanced study; a gazetteer lists 119 geologically interesting places, with notes on local minerals and gemstones and an index lists the geological terms printed in *italics*. The standard world-wide names of the geological time-periods used in the book are shown on the back cover.

The strategy underlying both book and programmes has attempted to help the newly interested amateur geologist to avoid the twin pitfalls of the 'local' and the 'chronological' approaches. The first real interest in rocks and the making of the landscapes often stirs on holiday, in some pleasant area where there is time to walk about and become familiar with the locality, to recognise the levels of the 'succession' of the rocks, and understand the events that produced them through the long span of geological time. After all this it can be a horrible disappointment to discover that moving a mere 50 miles away means that all the rocks are different and practically none of your detailed knowledge is relevant.

At this point the spark of interest can easily be extinguished. It is just as threatened by the sheer drudgery of trying to sort out the geological history of Britain by starting at the 'beginning' and working through the great periods of the classification. Because these periods embrace tens of millions of years and because Britain, in the shape we recognise is a comparatively recent phenomenon (it has for instance spent a large part of its history under water) the complexity of this approach is likely to overwhelm the amateur student.

On the Rocks offers another approach – that of processes; how the igneous and volcanic rocks are made from materials brought up from far below the earth's surface; how these rocks are worn down and transported by weather and water to form sedimentary layers; how seas flooding over the land leave great deposits of biological limestones; how the great overturning movements of sea-floor spreading and destruction fold and distort rocks by heat and pressure. These are some of the repeating sequences of geological history. If you understand them and recognise some of their effects you can apply your knowledge wherever you happen to be.

Any understanding of the processes that have affected Britain, a tiny fragment of the earth's crust, has to be based on ideas about the structure and behaviour of the earth. This book envisages the earth as a slightly distorted sphere, approximately 4,600 million years old, the same age as our own solar system, with a non-uniform interior. During its development it has separated into layers or spherical shells. In the middle is a *core*, probably of nickel iron and around that a layer of dense iron and magnesium silicate rich rocks called the *mantle*. Floating on the mantle is the crust – 5 kilometres thick under the oceans and between 30 and 70 kilometres thick under the continents – and always in motion.

It is part of the nature of deprivation that its victims are not aware of their plight. So it is not until you reach the happy state of being able to understand the landscape and 'read' the rocks that you realise what you have been missing. Without knowing it, the way we respond to landscape, owes a great deal to painting and literature, especially to the Romantic Revival that held sway in the early years of the 19th century, at the very time in 1807, when the British Geological Society first opened its doors.

Perhaps it is now time to let the upstart science enrich our vision.

Brenda Horsfield

VOLCANOES

1

Out of the fire

Volcanoes make a good beginning to geology. They are not only the most spectacular of all the earth's processes but also the most primitive. Day one in the life of a rock on the surface of the earth starts when it is erupted from a volcano. Volcanoes are powerful and irrational, they may sleep for hundreds of years before bursting into life again – little wonder they became Gods to the people who lived in their shadow.

Our ideal volcano is the perfect conical peak like Mt Fujiyama in Japan (fig. 1.1) formed from material brought up and thrown out through a single hole, or *vent*. Volcanoes are the product of processes going on deep down in the earth's mantle, 30-100km underground. The mantle rock may, under special circumstances, melt to produce some liquid rock, or *magma*. This is lighter and more mobile than its parent and so tries to escape to the surface of the earth. The volcano forms where this stream issues out. A volcano is, however, a curiously self defeating mechanism. As magma 'freezes' when the temperature falls below 800-1000°C the liquid stream fails to travel very far at the surface. The frozen magma piles up into a volcanic mountain which lies like a giant stopper, effectively blocking the vent. To avoid the rock pile, new escape routes may form that

1.1 The classic volcano shape – woodcut of Mt Fuji, Japan

grow new volcanoes as parasites on the flanks of the main one.

The differences between the shapes of volcanoes (fig. 1.2) are caused by variations in the properties of the magma. Most important is the ease with which the magma will flow, known as the viscosity. (A high viscosity means the liquid flows slowly: pitch is more viscous than treacle which is more viscous than water.) A highly viscous magma will tend to remain close to the volcanic vent and thus makes a steep sided cone. If the magma flows easily it will run downhill to solidify at a great distance from the vent and make a wide gentle dome, as in Hawaii. The hotter a magma is, the less viscous it becomes. Differences in composition may also be important.

Ultimately a volcano may start to control itself. The higher it grows, the more difficult it is for the magma to escape and so the magma cools, becomes more viscous and so makes an even steeper sided cone. Eventually the great weight of the volcano may block

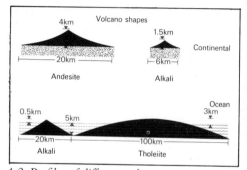

1.2 *Profiles of different volcano types*

off the magma provoking a battle between the gases and magma trying to escape and the volcano sitting on top, the resultant explosion may blow off the whole top of the mountain. It was such a battle that in 1883 sent a large portion of the island of Krakatoa, in the East Indies, up into the atmosphere in a gigantic explosion that was heard 4800km away in Mauritius and that rained the largest fragments over an area of four million square kilometres, leaving the dust in the upper atmosphere to turn many a Victorian sunset red.

Cont'd on page 10.

1.3 *World distribution of volcanoes*

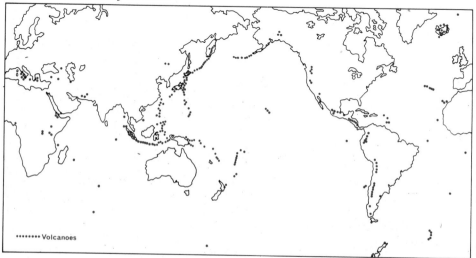

On the world map (fig. 1.3) it can be seen that volcanoes are restricted to narrow belts, the most important of which is the one running around the Pacific Ocean – the 'ring of fire'. To understand why volcanoes are so restricted we must study how and why the magma is formed. If the *Upper Mantle*, below the crust, contained magma everywhere, then there would be volcanoes everywhere. We know, from measuring the amount of heat coming out of the earth and from the construction of the *geothermal gradient* (the increase of temperature with depth), that the temperatures within the Upper Mantle are in general not high enough to melt the mantle rocks. As one goes to deeper levels the earth becomes hotter; however the temperature at which rocks melt also increases (see fig. 1.4).

If some mantle material from deep down rises, through convection, to nearer the surface, it can be hotter than the geothermal gradient and can melt. That there are such deep level mantle movements we know from the wanderings of pieces of the crust (see Chapter 3). The mantle rocks can flow, like a liquid of enormously high viscosity. All the volcanoes of basaltic composition that we see today, seem to be above places where the mantle is rising. Where two crustal plates are separating, material must rise from below to fill up the gap. The basalts on Hawaii and those in Africa lie above more isolated 'thermals'. The basalt magmas have been divided in two on their chemistry into the *alkali basalts* that are richer in alkalies and poorer in silica than the *tholeiitic basalts*. Volcanic provinces and volcanoes of alkali basalt tend to be smaller than those of tholeiitic basalt that make up such giant islands as Hawaii and Iceland. The differences between them seem to come

from the amount of melting. For small amounts, possibly less than 10% of the mantle rock, the melt or magma will tend to be alkali rich; for greater than about 10% it will be tholeiitic. Large amounts of mantle movement at the mid-ocean spreading ridge give tholeiitic magmas (Hawaii is exceptional not lying on a spreading ridge). Alkali basalts occur on small ocean islands and along rift valleys, as in East Africa, where mantle movements are restricted.

The andesite volcanoes that one finds above the subduction zone of the descending plate (see Chapter 3) are a little different in origin, forming from some mixture of melting of the mantle rocks with those of the sinking ocean crust. The presence of water may lower the temperature of melting. Andesite is a lower temperature, more silica rich magma than basalt, that tends to form the largest most perfect conical mountains, being more viscous than a tholeiite and occuring in larger volumes than an alkali basalt.

The names and chemical variations of these volcanic products, with the *plutonic* equivalents, are listed overpage (fig. 1.5).

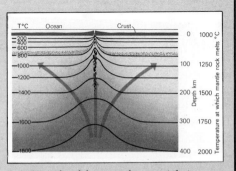

1.4 Deep level hot mantle material rises to shallower levels where it can melt to form a magma

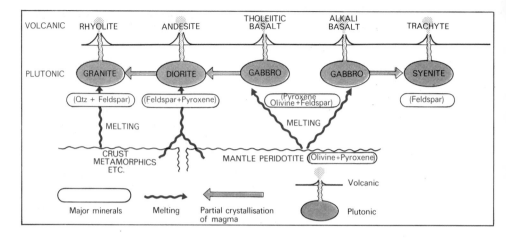

VOLCANIC	RHYOLITE	ANDESITE	THOLEIITIC BASALT	ALKALI BASALT	TRACHYTE
PLUTONIC	GRANITE	DIORITE	GABBRO	GABBRO	SYENITE
	(Qtz + Feldspar)	(Feldspar+Pyroxene)	(Pyroxene Olivine + Feldspar)		(Feldspar)
	MELTING			MELTING	
	CRUST METAMORPHICS ETC.		MANTLE PERIDOTITE (Olivine+Pyroxene)		

Major minerals   Melting   Partial crystallisation of magma   Volcanic   Plutonic

'Of course, it could never happen here!' Volcanoes are not going to start appearing in an English back-garden. However, in the past, as we shall see, the most devastating volcanic eruptions have been a familiar part of the scene. Volcanic ash and *lavas* weather quickly. The minerals within the rocks were formed at very high temperatures from the magma and at normal temperatures they are unstable in the presence of water and break down. This is why volcanic products make such a good fertile soil. It also means that the form of a volcano can be lost after a relatively short (in geological terms) period of time. The last volcano of the Massif Centrale in France stopped erupting only 7,500 years ago, yet already the cones are starting to lose their shape. After a million years, there would be a few small, well rounded hills. The last British one gave a final volcanic belch some 50 million years ago. Although the conical shape has everywhere been lost, it is possible, in some places, to see the central vent or *'neck'* of the volcano, that contained hard rocks, protruding from the softer surround. The rock of the neck of a

volcano may be as hard as concrete and may remain as a pillar as at Ship Rock, New Mexico (fig. 1.6) or to a lesser extent in some of the similar, but much older rocks of South Scotland.

Around Britain it is always more common to find the lavas and ashes that came out of volcanoes, rather than the volcano itself. However, the old eroded volcanoes of Britain have been the centre of some of the studies to see just how volcanoes work, because through them we can see 'inside' a volcano. They have preserved, in cross section, many of the different events that followed one another over a long period of time in the history of one volcano. Without the threat of being blown to bits or smothered with gas and ashes we see the volcano as a fossil; the eruption and devastation can safely be left to the imagination.

One of the most familiar of these former volcanoes lies in the centre of Edinburgh. Both Castle Rock and Arthur's Seat are remnants of a volcano that died about 350 million years ago (fig. 1.7) during the early part of the Carboniferous Period (see back cover). Both these rugged hills

1.5 Origin and nomenclature of the igneous rocks (left)
1.6 Ship Rock, New Mexico – volcanic plug resisting erosion (right)
1.7 Arthur's Seat volcano (below)

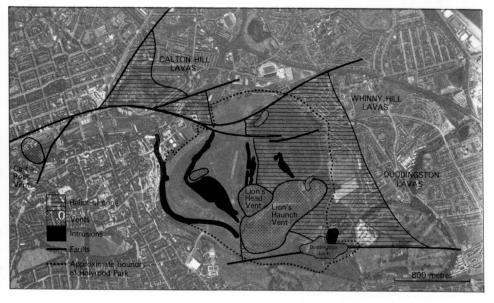

CALTON HILL
LAVAS

WHINNY HILL
LAVAS

DUDDINGSTON
LAVAS

Castle
Rock
Vent

Lion's
Head
Vent

Lion's
Haunch
Vent

Duddingston
Loch

Relics of cone

Vents

Intrusions

Faults

Approximate boundary
of Holyrood Park

800 metres

exist because they have resisted erosion. The former volcano is left only as tilted fragments, and so piecing together its history requires some geological detective work. All that remains are the edges of the cone, five vents or necks up which the magma travelled and some *intrusions* in which the magma, instead of erupting, became squeezed into pockets between pre-existing rocks.

The cone was made of both volcanic ashes and lavas. At the edges of the cone these layers are thin but they thicken rapidly towards the vent from which they were erupted. Thirteen distinct episodes of lava flows have been traced on Whinny Hill with many ash layers inbetween. These can all be matched with the lavas and ashes of other localities to help reconstruct the eruptions and find from which vent each one was launched(fig. 1.8).The two large vents on Arthur's Seat were infilled with small intrusions and with a rock called *agglomerate* that is characteristic of volcanic vents, being a mixture of volcanic blocks with rocks carried from below the volcano, all forced together by the explosive power of the eruptions. Sometimes the magma formed intrusions by forcing two earlier layers apart and filling up the space between to form a *sill*, that has sometimes reached 25m in thickness. These sills can be distinguished from the normal lava flows because only the sills burned the rocks already present above them (fig. 1.9). This burning has baked or even melted the earlier rocks.

It is not clear how the vent that underlies Edinburgh Castle fits into the pattern of the remains of the main part of the cone further to the east. The Arthur's Seat volcano probably erupted for a hundred thousand years or so and may have stood a thousand metres up above surrounding lagoons – high for our present day mountains but a mere nipper among volcanoes (fig. 1.10). (This is typical of its magma type – an alkali basalt.) During this period there were many other similar volcanoes active across the Central Lowlands and Southern Uplands of Scotland. Further to the west, around the Clyde Plateau the lava flows cover an area of 7,500 square km. Some of the other vents have also been exposed by recent (mostly glacial) erosion. Perhaps the most spectacular is the Bass Rock, an island surrounded by sheer cliffs rising 100m from the Firth of Forth (fig. 1.11).

If we travel 150km to the N.W. and forward through 280 million years in time we can find remnants of some more massive volcanoes. During the early Tertiary Period, some 70 million years ago, the West of Scotland, the Hebridean Islands and Antrim were all the site of major volcanic activity (see Colour 1). The earliest lavas (of alkali basalts composition) were erupted from fissures and piled up to form high flat lying plateau mountains. In the Island of Mull there still remain some 2,000m thickness of these. At one time these plateau lavas covered an enormous area from the Hebrides across Northern Ireland where 3,500 square km still remain. Each flow was fed from a new split in the crust – a kind of volcanic eruption that is only possible where the

1.8 Cross section of a volcano

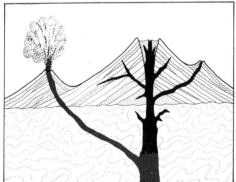

1.9 Cross section of a sill

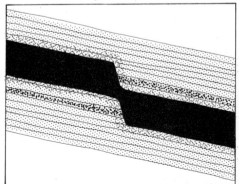

whole region is in tension (see Colour 1). These fissures, once filled with magma, cool to form sheet-like intrusions known as *dykes*. The flat lying lavas were about 15m thick, each covering a very wide area. After each flow, the surface of the lava became slightly weathered to form a fertile red soil before the next lava flowed over the top. These soils provide planes of weakness in the hills and cliffs that take on a characteristic *trap* topography; trap being the step-shaped profile (fig. 1.12).

The second stage of activity saw a change from *fissure eruptions* to enormous volcanoes. The eroded remnants of these former mountains can be seen at Skye, Rhum, Ardnamurchan, Mull, and Arran on a line that must mark some deep level crustal fracture; with three in S.E. Northern Ireland and extraordinarly, 350km away, one at the island of Lundy off the North Devon coast. All reveal some of the life history of the volcano; the largest ones show that

1.10 *Reconstruction of the Carboniferous Arthur's Seat Volcano*

1.11 *Bass Rock, Firth of Forth — eroded neck of a volcano*

1.12 *Trap topography, Central Scotland*

In many areas the slow cooling of the thick lavas has formed *columnar jointing*; the best known examples being at the Giant's Causeway in Antrim (fig. 1.13) and at Fingal's Cave, on Staffa in the Hebrides. The rock face resembles a huge surreal stack of pencils, polygonal columns of stone like giant crystals, each one 20-80cm across.

In cooling and crystallising the magma loses volume. When the lava flow (or sill) cools it begins to solidify from the top surface, for that is where it can lose its heat. The most economical way for it to contract with the shortest length of crack that is always close to a shrinking centre, is for it to form a hexagonal mesh (fig. 1.14). Once these

1.13 Giant's Causeway, Antrim – columnar jointing

polygonal cracks form (like the cracks that form in mud as it dries) they will continue to follow the solidification through the magma. If the cooling surface, for some reason, turns a corner, the columns will turn corners to follow it. Where the ground beneath a flow is also cold, columns may grow from below to meet the ones forming from above.

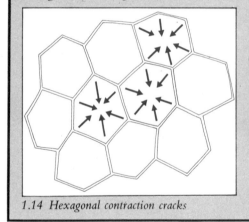

1.14 Hexagonal contraction cracks

when the volcano reaches a certain size (3-4,000m high) the magma, instead of erupting, builds up inside the mountain to form a *magma chamber*. The cone of many of the large volcanoes seems to be very plastic and one of the best ways for predicting eruptions on active volcanoes is to watch for changes in shape of the cone as these magma chambers start to fill or empty.

The composition of the magma may begin to change as crystals form and sink to the bottom of the chamber taking out magnesium and calcium faster than silicon, sodium or potassium and thus making the composition

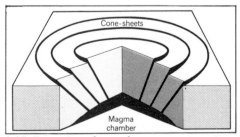

1.15 Formation of cone sheet magma intrusions

1.16 *Wizard Island rising inside Caldera (collapsed volcano) at Crater Lake, Oregon*

move towards a rhyolite (see fig. 1.5). The gas pressure builds up as the amount of liquid in which the gas was originally dissolved under pressure (like carbon dioxide in a soda syphon) decreases. The rising gas pressure may force up the top part of the volcanic cone, like a bottle stopper, and magma may intrude along this zone of weakness to form a conical dyke known as a *cone sheet* (fig. 1.15). Eventually the build up of gas pressure may become critical and the contents of the magma chamber will burst out

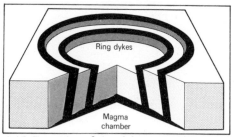

1.17 *Formation of ring dyke magma intrusions*

in a catastrophic eruption. Hundreds of kilometres away across the whole North Sea there are up to 30m of ash deposits that all came from the eruptions on the other side of Scotland! After the eruption the empty chamber will collapse bringing down with it the whole top of the mountain. The depression left behind is known as a *caldera*. The most perfect example is at Crater Lake, Oregon (fig. 1.16) but the best known eruptions of this kind were at Santorini (around 1400 BC) and at Krakatoa. In both cases the caldera is obscured by the sea. During this collapse a different kind of stopper may form that cones inwards rather than outwards. The gap between the inner and the outer wall may be filled with some remnant magma to form another kind of circular intrusion – the *ring dyke* (fig. 1.17). In cross section these dykes form rings like those on a tree – each ring marking a phase of a former eruption.

At both Mull and Ardnamurchan there are well over one hundred cone sheets, each 3 or 4m thick, in combination with about ten ring dykes of much greater thickness – each the result of a major catastrophic eruption with subsequent collapse. In both volcanoes the original vent became so blocked that a new one was tunnelled out a short distance from the first, making a new cone on the side of the old one. At Ardnamurchan the volcano shifted on two such occasions, each time taking its cone sheets and ring dykes with it (fig. 1.18).

At Skye and Rhum the volcano has been eroded even deeper to reveal the magma chamber itself, in which one can trace the cycles of crystallisation and eruption, followed by the arrival of more magma, as layers within the intrusion. Magma erupted from these chambers is often full of big crystals, known as *phenocrysts*. The rock texture as a whole, of big crystals in a fine grained material, is known as *porphyritic*. Phenocrysts may be easily mistaken for other kinds of crystals in lavas. Gas bubbles, known as *vesicles*, may become partially infilled with crystals to become *amygdales*, that keep the bubble's oval shape but form after the lava has solidified. Some alkali basalts carry small aggregates of crystals, called *xenocrysts*, that have risen with the magma from where it originated in the mantle. These are commonly of the pale green mineral olivine. These xenocrysts should not occur with phenocrysts because a magma that has a chance to crystallise will drop any fragments of the parent mantle rock that it was carrying.

After these volcanoes became extinct the region went back to a stage of tensional activity with dykes oriented N.W.-S.E. Around Mull and Arran, the crust has increased in length by

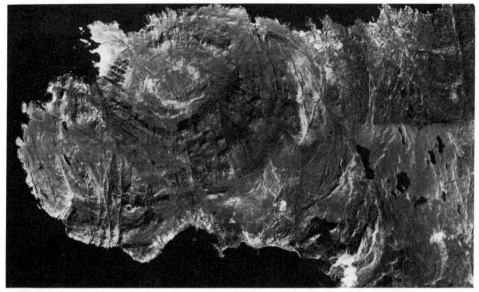

1.18 *Aerial view of ring dykes and cone sheets of the eroded remnant of the Tertiary Ardnamurchan (Scotland) volcano*

between 5 and 10% during this basaltic dyke emplacement. Some of the most widely travelled cut across northern England to die away on the Yorkshire coast and one or two run through North Wales and the N.W. Midlands.

However, the most extensive of the British *sheet intrusions* is much older. The Whin Sill (tholeiitic basalt) is of early Permian age and covers an area of 2,500 square km in Northern England (see Colour 2) averaging about 30m in thickness and running inside, and parallel to, the bedding planes of the Carboniferous sediments. At outcrops it is tougher than the surrounding material and forms cliffs along the Northumberland Coast, the crags on which the central section of Hadrian's Wall was built and, further south, the lips of some of the Teesdale waterfalls (fig. 1.19).

1.19 High Force, Teesdale – outcrop of the Whin Sill (tholeiitic basalt)

1.20 British Pillow lavas

Greater volumes of volcanic rocks formed still earlier in the British geological history where island arcs, like those now found festooned around the Pacific, extended across the Lake District and Wales, forming 3,000–4,000m of andesitic lavas and *tuffs*, ash deposits now turned into hard rock. Most of the volcanoes are missing, only their products survive. In some of the lavas of this period (Ordovician) in Southern Scotland the rock has taken on a pillow-like texture (fig. 1.20).

Although we may be thankful that all the volcanoes, once active in Britain, are long extinct, their lavas and eroded remains still provide some of the grandest scenery.

> *Pillow lavas* have recently been seen forming in Hawaii. They develop where the magma is issuing underwater. Small bursts of magma become quickly frozen around the edges in the shape of a large bubble about a metre across. As the magma bursts out, a new bubble will form on top. The weight of these 'bubbles' of magma on top of each other causes flattening that makes the pillow structures. They can also be seen in basalts at Anglesey and in Devon.

2

SEDIMENTS

Break and make

In this chapter we are going to look at a simpler series of rocks: *sediments*. These are rocks laid down after travel as particles on the wind or in the water. The processes involved can be seen in the weathering and deposition taking place all around the country today. It cannot have been long before the 'first' volcanic rocks began to weather, to form sediments, the 'second' rocks.

To follow the story of sediments we must first look at this weathering and disintegration of the pre-existing rocks. Not only can we see all the processes taking place on our doorstep; erosion may even be affecting our doorstep. Wherever there is rock exposed, there is weathering: on the screes of Wastwater in Cumbria (fig. 2.1); on the buildings of Oxford where acid in the rain is dissolving away the stone (fig. 2.2); and on cliffs all around the country where sea-side houses in Yorkshire, Hampshire and Norfolk (fig. 2.3) are undercut by the action of the waves and have to be abandoned. During erosion, rocks suffer both a chemical and a physical attack from water. Some soluble elements such as sodium will dissolve. Many of the minerals that have formed in rocks at high temperatures are unstable at low temperatures in the presence of water, and will breakdown to form new minerals (hydrous silicates also known as *clay minerals*). These low temperature

2.1 Erosion: Wastwater screes, Cumbria

minerals form clusters of very small crystals in place of the original grain. The only common mineral that can form at high temperates and yet remain stable at low temperatures in the presence of water, is quartz (chemical formula SiO_2-silica). The initial shattering of a rock may break off fragments containing many grains but further disintegration separates the rock particles into the individual mineral grains. If left to itself, this is the process by which a rock can become a soil. In volcanic regions the soil is very fertile because the minerals of the rock are unstable and will break down fast to release many elements that are absorbed by the plants before they can be washed away in streams. Few soils are future sedimentary rocks, for continuing erosion will wash them away. In more mountainous areas,

2.2 and 2.3 Erosion: Emperors heads outside the Sheldonian, Oxford (right) and coast south of Lowestoft (below)

where there are exposed rock faces, rock fragments, rather than a soil, are carried away by the river. Rarely, it is possible for a scree to become turned into a rock in its own right (fig. 2.4). This is termed a *breccia*, a rock composed of angular fragments of other rocks. Once these fragments have been transported in a river for a short distance, they start to lose their sharp edges through wearing against one another. The rock particles are not actually supported in moving water unless they are small enough to be *colloidal*. (When particles are smaller than dust size they may be suspended in the water through the interaction of tiny electrostatic charges. Milk is a

2.4 A breccia

The process whereby a soft clay, a sand dune, or a pebbly beach become turned into hard rock formations is known as *diagenesis* or *lithification*. Most of the changes take place as a large weight of overlying sediments builds up to squeeze the individual particles together. As the water in the original sediment is forced out, the particles are forced to adapt their shapes to fit against one another. Through the formation of some new mineral grains that dissolve out from the water, the sediment is bonded together to become a rock. A volcanic ash may become lithified into a *tuff* just as a clay becomes a *shale*, or mudstone.

a river follows curves or meanders (fig. 2.5), the faster water on the outer part of the bend will erode, while on the inside, the slower water will deposit. In times of flood the river may rise above its ordinary banks and flow over the surrounding land, but as the river can flow much faster in the original deeper channel than it can on the flooded land, it will tend to deposit much of its load on the banks, where the speed suddenly slows. This builds up the banks to form natural *levées* that are the river's own form of flood control. The deposition of sediments

colloid.) Larger particles are rolled along by the current crashing one fragment against another. Rocks become pebbles as their surfaces become smoothed. Continued battering grinds down the size, making pebbles into gravel and gravel into sand.

Deposition of particles will occur where the speed of the river becomes too slow. Where the river gradient flattens, the heavier part of the load will fail to move any further. So where

2.5 A meander in the River Wye

2.6 Britain's outline is considerably changed with added deltas

2.7 A small recent delta, River Romesdale, Isle of Skye

on the flood plain enriches the soil. However, as with living on the side of the volcano, anyone who reaps the rewards of farming the flood plain cannot avoid reaping the occasional flood. As the Egyptians have found in taming the Nile floods with the Aswan Dam, it does not take long before the fertile farmland deteriorates without its annual dose of nutrients.

The most marked change in velocity of the river and the most likely place for fresh-water sediments to build up and be turned into a rock, is at the *delta*, where a river enters a lake or the sea. The delta shape is a result of the sudden deposition of all the material being carried by the river as it enters still water. In the geological record deltas are relatively common and yet around the coasts of Britain today, all the major rivers end in long V-shaped marine estuaries that give our familiar 'deflated' profile. This

is because for a long period during the recent Ice Age, only a few thousand years ago (see Chapter 9) the sea level was considerably lower, and the rivers had to carve for themselves deep valleys down to reach it. The estuaries are the flooded and partly silted remnants of these valleys. Estuaries make bad examples of sediment deposition as they tend to funnel the tides, making very strong currents that carry sediment away, though in time they will fill up to become deltas. Other rivers, the Nile, the Mississippi and the Rhine for instance, have already infilled their Ice Age estuaries, in the space of only a few thousand years. If left alone for a few million years, such deltas would expand to cover large areas of continental shelf (fig. 2.6); for instance the Rhine would join with the Thames to cover a large part of the Southern North Sea (as was the case only a few million years in the past). We can still see the form of deltas where streams bring mountain debris down to the shore (fig. 2.7). The fan shape resembles the Greek letter Delta Δ – hence the name.

Deltas are a mixture of swampy ground, that is the flood plain of clay and mud deposition, laced with river channels of deeper moving water lined with sand. As the sand builds up the river will carve for itself a new channel through the muds (fig. 2.8). This delta pattern of mixed sediments may build up considerable thicknesses of rocks when there is a balance between the sinking land and the sediment supply.

Such conditions have existed during several periods in the building of Britain. The most important of these was in the Upper Carboniferous Period when deltas extended over much of Northern England forming the 'Millstone Grit' (up to 2,000m of *sandstones* and shales) and subsequently the 'Coal Measures' (see Chapter 8). Deltas existed in Triassic times from rivers flowing northwards off the Welsh and Midland uplands carrying sand and gravel onto a plain of muddy lake flats; in the Middle Jurassic Period when there were deltas over much of Yorkshire; and in the Cretaceous Wealden of Sussex where rivers again brought debris off the landmass to the north. Still more recently during the lower Tertiary, the Thames Valley Region was a delta that advanced on three successive occasions into the sea

2.9 Chesil Beach, Dorset with drowned landscape behind

to the east. On each occasion, it was submerged by advancing marine conditions caused by the rapid sinking of the North Sea basin.

At the boundary of the land and the sea is another sedimentary environment – the beach. Many of the beaches around Britain have, like the estuaries, arrived from sea level changes connected with the end of the last ice advances. As the sea rose, the beach has been trundled along by the breaking waves, ahead of the rising tides and has often built up as a bank or reef some distance away from where the natural coastline should be. Such shingle bars with lagoons trapped behind are found at Chesil Beach in Dorset (fig. 2.9)

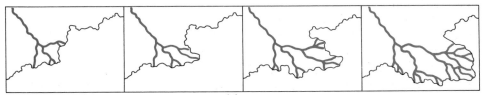

2.8 Pattern of changing river course across a delta

and have blocked off rivers at Newhaven and Orford in Suffolk. At Dungeness huge expanses of shingle are still being added to, from fossil beaches left submerged further out in the Channel. Along the USA East Coast these reef beaches have trapped enormous expanses of swamp behind them (fig. 2.10).

A beach normally reflects the kind of material available from the local rocks. Around the south and east coasts of England the chalk erodes fast leaving behind the hard wearing flint nodules to form great banks of shingle. Elsewhere around Britain, sand or mud beaches are more common, as the rocks more easily break down to smaller sized fragments. The shingle beaches turn up in the British rocks as *basal conglomerates* (a conglomerate is a rock made up of rounded pebbles cemented together) that record where the sea slowly rose to cover over the land (fig. 2.11). Basal conglomerates are common at the bottom of many marine sedimentary sequences.

The overall currents of an area, as with the river, determine whether mud or sand is deposited on, or eroded from, the beach. In general the lighter fine grained material is carried away to slower moving marine environments. Such muds may build up over a very wide area – such as the North Sea today, that is receiving a fine sediment input from a number of rivers. The more finely grained the mud, the slower it will tend to accumulate.

Where sediments build up on the edge of a steep slope, perhaps the front of a delta or at the continental shelf edge, there may be catastrophic slips, like mudslides or avalanches, in which a whole mass of sediments takes off

2.10 *Satellite view of reef beaches in eastern USA. Long Island and New Jersey*

2.11 *A conglomerate*

and may keep rolling down quite shallow inclines at up to 80km per hour. Small ones can be simulated in an experimental water tank (fig. 2.12). They are known as *turbidity currents*; the largest one in recent times formed

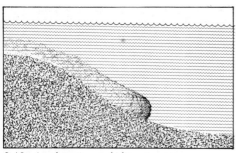

2.12 *A submarine turbidity current*

after an earthquake on the Grand Banks, Newfoundland in 1929. An enormous sediment avalanche was set off in an underwater canyon that rolled down across the floor of the Atlantic, still breaking all the telephone cables in its path 13 hours later, by which time it had travelled some 600km! Recent sediment cores in the area show that more than a metre of sediment was deposited over a huge area of the sea bottom. This sediment is known as a *turbidite*. As the particles within the turbidity current settle out, the larger ones sink to the bottom, giving a natural sorting process known as *graded bedding* (fig. 2.13), that is the characteristic feature of the turbidite. Where a rapidly sinking trench forms in the sea floor, it will tend to fill up with turbidites and other sediments containing a mixture of particle sizes and particle types – all known under the title of *greywacke*. Greywackes may form enormous thicknesses, as in the troughs that developed adjacent to newly uplifted mountains in parts of Northern Wales and the Southern Uplands of Scotland some 450 million years ago.

Sediment structures, such as graded bedding, are important in unravelling the story of the conditions when the

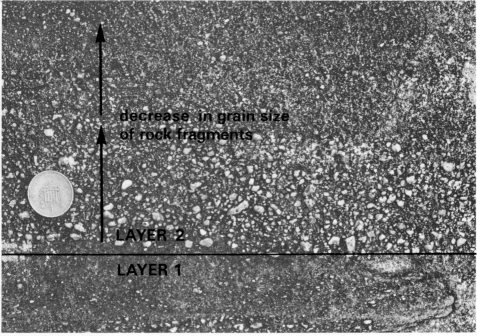

decrease in grain size of rock fragments

LAYER 2

LAYER 1

2.13 *Graded bedding*

2.14 *Fossil ripple marks in sandstone*

2.16 *Fossil dinosaur footprints*

2.15 *Fossil mudcracks in mudstone*

sediments were formed. River currents tend to push the sand into mounds or sand waves. The former positions of the front ends of the sand waves can be seen in *cross bedding* – each one is traced by slight differences in the composition of the material deposited. From these traces it is possible to tell the direction of current flow at the time of deposition. Smaller scale sand waves may form ripples like those on the sea shore, that may also be preserved in the rock (fig. 2.14). Muds that dried out to form mudcracks tell of drought (fig. 2.15); or one may find traces of life; fossils or even footprints (fig. 2.16). The largest scale features are the cross bedding formed from

2.17 Wind blown dune cross bedding in sandstones

windblown sand dunes, each of which may be ten or twenty metres high (fig. 2.17). When preserved as sandstones they reveal the directions of the prevailing winds at a particular period. Windblown sands and desert environments were very important during two periods in the building of Britain: the Devonian and the Permian. More recent windblown deposits across Europe and Asia are made up of *loess*, or windblown dust, formed out of dried mud deposited by glaciers. Dust storms that carry both sand and dust may travel enormous distances; there are even sand dunes on the Canary Isles that have come from the Sahara! In many parts of the world it is only the vegetation that binds the soil and prevents it from being blown away. Before plants covered the land during the Devonian Period, all the continents must have been dust bowls of rapid erosion and shifting dunes.

Under the microscope these sands of the desert can be distinguished from the sands of the beach or the sands of the river (fig. 2.18). Because the particles in the air move much faster than those in the water, they cause much more rounding and polishing when they collide. Also, the longer the particles have been transported, the more worn they become. Thus the grains of a beach sand become much more rounded than those of the river. Sometimes, with a change in climate or geography, a desert may become washed away. The pre-Cambrian

Cont'd on page 30.

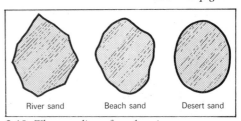

| River sand | Beach sand | Desert sand |

2.18 The rounding of sand grains

Structural Geology

The horizontal strata common in many sedimentary rocks provide useful markers for seeing the effects of movement and distortion. *Structural geology* involves unravelling the complexities of rocks that have been tilted, bent or broken.

The slant of a rock series is described by its *dip* and *strike*. The dip measures the steepest tilt of the plane, and is perpendicular to the strike that measures the orientation in terms of the compass of the horizontal line within the plane (see fig. 2.19).

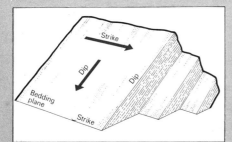

2.19 Tilted rock layers

A break in the rock series is a *joint*. If there has been movement along this break it is a *fault*. The *fault plane* possesses its own dip and strike. The movement that has occurred along the fault is referred to as the *throw*. Unless one can find an exact marker to match on either side, the throw can be measured either along the dip or along the strike of the fault plane. (If the strata are horizontal and the movement is also horizontal it may be impossible to detect (fig. 2.20). Faults were traditionally classed as: *normal*, in which one side has dropped down in relation to the other; *thrust*, in which one side has been pushed over the other; and *strike-slip*, in which the movement was horizontal. Intermediate directions of movement and even variable movement are common along many fault planes. On small faults, scratch marks may develop that reveal the actual direction of movement. These are known as *slickensides* (fig. 2.21).

Many rock formations behave plastically under stress and the rocks fold. Folds are divided into: *synclines*, when the rocks are bent through a trough or 'U'; and *anticlines* when they are folded through an arch or '∩'. A *monocline* is when one side of the fold is horizontal (fig. 2.22). The sides of the fold are known as the *limbs*. The plane that bisects the angle between the two limbs is the *axial plane*. The line of intersection where the *axial plane* meets a *bedding plane* is the *hinge line* that gives the direction or strike of the *axis* of the fold at that point. If the *axial plane* is vertical and the *axis* horizontal (the simplest case) the fold is said to be upright and symmetrical (fig. 2.23). If the axial plane is inclined, then the fold may be a *monocline* or an *overfold* or a

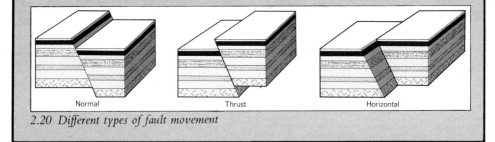

2.20 Different types of fault movement

2.21 *Slickensides from a fault plane showing movement of overlying rock*

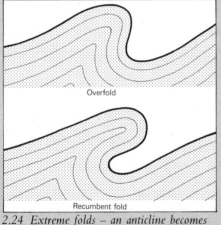

Overfold

Recumbent fold

2.24 *Extreme folds – an anticline becomes pushed towards a 'nappe' (see fig. 3.11)*

Monocline's

Anticline

Syncline

2.22 *Folded rock layers*

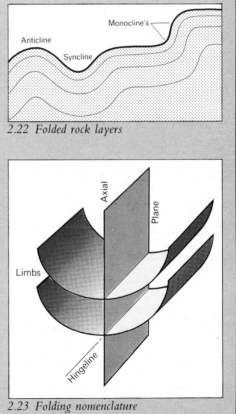

Axial

Plane

Limbs

Hingeline

2.23 *Folding nomenclature*

recumbent fold (fig. 2.24). If the axis of the *axial plane* is inclined, then the fold has *plunge* (fig. 2.26), that is a measure of the dip of the axis. Folds may get extremely complicated and confused (fig. 2.25). Different layers of a series of rocks may behave in completely different ways. Small folds in rocks may be part of much larger folds when one episode of folding is superimposed on another.

2.25 Complicated folding

The structural geologist's work is to study an area of country and to map all

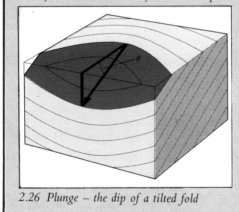

2.26 Plunge – the dip of a tilted fold

the information of strikes and dips etc. to obtain the shapes of the structures. Different colours may be used on a geological map to mark the different ages of rocks. The shapes of the valleys and mountains of the countryside will impose their own patterns on those of the rock structures, making the job of visualising the folds and faults one requiring good skills of 3D perception.

Now we can ask the question 'why?' Why did the rocks become folded and broken? Why are there mountains in Scotland but not in East Anglia. Why does Britain lie on the edge of a vast ocean?

Torridonian sandstones of Scotland have the roundness of desert sands and yet the cross–bedding of a river delta.

The processes of erosion, transportation and deposition are the visible part of the crustal rock cycle. In the next few chapters we are going to look at some of the hidden parts of the journey, that take place deep underground, before the rock can return to outcrop at the surface and so start the cycles all over again.

TECTONICS

3

United Kingdoms

Within the last fifteen years a revolution has taken place within the study of geology. The earth's crust is now known not to be a fixed shell, but more like a series of 'fish scales' that move in relation to one another. New ocean floor crust is created where two of these 'scales' move apart. Old ocean crust is destroyed where one 'scale' moves beneath another, back into the mantle. These cycles are permitted because the crust underlying the world's oceans is moving and forms a kind of conveyor belt system upon which the lighter 'scales' of continental crust float. These 'scales' are known as plates but because they lie on the surface of the spherical earth they are actually more saucer shaped. They are rigid, generally 70-80km in thickness and include both the crust and the top section of the Upper Mantle.

The study of the way these *plates* move in relation to one another is known as *Plate Tectonics*. (Tectonics means building) (fig. 3.1). Most volcanoes (see fig. 1.3), mountain buildings and earthquakes happen at the boundaries of these plates. Plate tectonics has helped to find order within many parts of geology, for many of the problems of both British and world geology were waiting for a grand design which would inter-relate them.

There are at present seven major plates on the globe (fig. 3.2). Between some of these there are smaller plates or platelets. Although plate movements are slow, a few centimetres each year, over geological periods of many millions of years this adds up to thousands of kilometres. Where a plate boundary has been carefully studied it is seen that movement is intermittent. For much of the length of the San Andreas fault in California, that marks the boundary between the Pacific and American plates, movement takes place as catastrophic earthquakes, such as the one at San Francisco in 1906 when the fault moved 2m (fig. 3.3). As average movement along the fault seems to be about 5cm each year, a similar earthquake should be expected every forty years. San Franciscans know that the longer they wait until the next one the bigger it is likely to be.

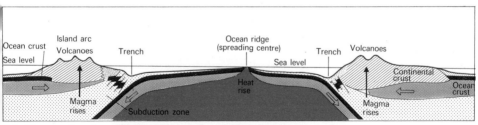

3.1 Cross section through the earth's crust and plate boundaries

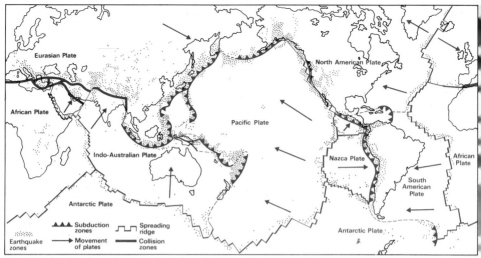

3.2 *World plate boundaries and earthquakes* (above)

3.3 *Aerial view of the San Andreas fault 'wound'* (right)

Central Iceland lies directly atop the mid-Atlantic spreading ridge but no splitting was recorded for the ten years up to 1976, when there was a renewed period of earthquakes and volcanic activity, with crustal widening, suggesting a new dyke was being formed along the split.

Most of the world's earthquakes occur along the greatest build-ups of stress at the plate boundaries; in fact these earthquakes are used to define the plate boundaries (fig. 3.2). Some earthquakes represent a new break forming through the rocks, but the majority are movement along an old wound, for once formed, these fault planes remain planes of weakness. Britain does not now lie near any plate boundary, but still suffers from minor earthquakes as there are still some slight re-arrangements of mountains and crustal blocks. The country is, however, meshed with old breakages or faults. These were formed in the past, when plate boundaries were closer or even passed through the country. Every one represents a former earthquake or a series of earthquakes. The major British faults and earthquakes are shown in fig. 3.4. The most destructive earthquake of recent

3.4 *British earthquakes and faults (with deformational structures in the Highlands)* (right)

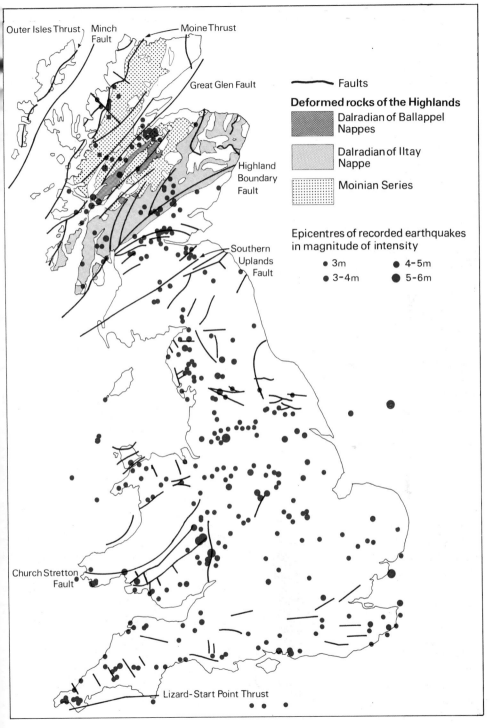

Outer Isles Thrust — Minch Fault — Moine Thrust

Great Glen Fault

Highland Boundary Fault

Southern Uplands Fault

Church Stretton Fault

Lizard-Start Point Thrust

Faults

Deformed rocks of the Highlands

Dalradian of Ballappel Nappes

Dalradian of Iltay Nappe

Moinian Series

Epicentres of recorded earthquakes in magnitude of intensity

● 3m   ● 4-5m
● 3-4m   ● 5-6m

centuries was at Colchester in 1884 (fig. 3.5). Many houses and churches were damaged and the 'quake was felt right across England. The old fault that moved lies hidden beneath more recent sedimentary cover.

EXTERIOR OF LANGENHOE CHURCH

INTERIOR OF LANGENHOE CHURCH

CONGREGATIONAL CHURCH, LION WALK, COLCHESTER
(The dotted portion of the Steeple was shaken down)

IN THE HYTHE, COLCHESTER : THE RUSH FOR THE GASWORKS

PELDON CHURCH

ROSE INN, PELDON

ON THE QUAY, WIVENHOE

COTTAGE AT ABBERTON

THE RECENT DISASTROUS EARTHQUAKE IN EAST ESSEX

3.5 Damage caused by the Colchester Earthquake 1884

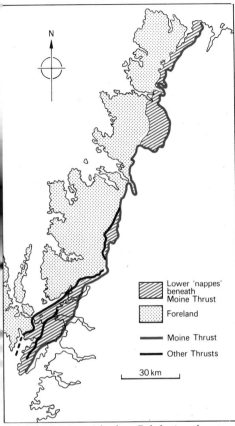

3.6 *The Moine and other Caledonian thrust faults of North West Scotland*

Map legend:
- Lower 'nappes' beneath Moine Thrust
- Foreland
- Moine Thrust
- Other Thrusts

30 km

3.7 *Cliff at outcrop of Moine Thrust*

The biggest British faults cut across Scotland. To the far North-West there is the '*Moine Thrust*' on which old continental crust was thrust over younger limestones of the Cambrian Age (fig. 3.6). Sediments of the 'Moine' that formed off the edge of a continent have been moved east next to the 'Torridonian' sediments that formed at the same time but on the continent itself. The thrust hardly slopes at all and the total movement is not known but thought to be at least 20 or 30km. The harder crystalline rocks resting on the younger sediments often form small cliffs (fig. 3.7). But the fault plane lies too flat to score a deep mark on the landscape.

To find a scar more like the San Andreas Fault, we have to travel 100km to the east to the *Great Glen Fault* (fig. 3.8). The scar is really not a single plane but a zone of movement about a kilometre wide in which the rocks from either side have become incorporated and broken up to form a kind of lubricant. (On the smallest scale the grinding action of one rock mass moving over another may form a kind of rock paste called a *milonite* that consists of tiny submicroscopic fragments of the original rock. The heat of friction on the fault surface can even cause the rock to start melting and form a glass.)

The material within the Great Glen Fault zone has weathered faster than the surrounding hard rocks and a series of ice-worn lakes has developed along it. Like the San Andreas Fault, this has been the site of large scale horizontal movements, but exactly how much and even in which direction is still a matter of considerable argument. Up until only a few years ago the matter seemed settled as it was thought that the Strontian Granite and the Foyers

3.8 *Aerial view of the Great Glen fault 'wound'*

Granite (see fig. 3.9) were split halves of the same intrusion that had been separated by 100km movement of the North of Scotland towards the S.W. But this matching has now been disputed. Alternatives that have been suggested include a movement of between 200–300km in the same direction, and even a movement of about 100km in the opposite direction, that would bring the north coast of Scotland parallel with the southern edge of the Moray Firth. At both ends of the fault it seems to fade out.

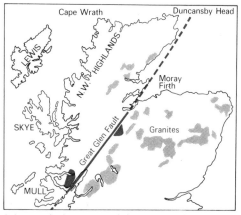

3.9 *Matched granites of the Great Glen Fault*

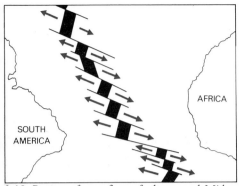

3.10 *Pattern of transform faults around Mid Atlantic spreading ridge*

Investigations of the geological structures through the Moray Firth during oil exploration failed to show any marked break in the strata, where it would be expected. But in many reconstructions the fault has been shown continuing through the Walls Fault across the centre of Shetland and even up to Spitsbergen more than 1,000km away, off the north coast of Norway. Near Inverness there has been vertical movement of more than 1.5km and even within the last two hundred years there have been sixty minor earthquakes along it. Clearly it is a major deep level crustal wound.

South of the Great Glen Fault and running almost parallel with it, is the *Highland Boundary Fault*, marked by the southern boundary of the mountains. An enormously thick pile (perhaps 7km) of Devonian sediments formed in a gradually subsiding trough to its south, when it behaved as a *normal fault*, but earlier in its history it seems to have operated as a *strike-slip fault*. In common with the Great Glen Fault, it seems just to fade away. At its south west end its continuation into the Isle of Arran, only a few kilometres off

the mainland, is missing. This kind of feature, of faults that end abruptly, is typical of the stepped arrangement of the *transform faults* around a mid-oceanic spreading ridge (fig. 3.10).

These three major fault displacements, each of which took place through many thousands of minor fault movements (each an earthquake), could only have taken place near plate boundaries. These faults are but one part of a reconstruction that demonstrates how Britain, before the Devonian, was split by an ocean, more or less along the England-Scotland border. Large scale movement is also recorded in the giant rock folds of the Scottish Highlands.

The largest is the *Iltay Nappe*. (A *nappe* is an *overthrust* in combination with a *recumbent fold* in which one rock mass may become forced to 'flow' over another (fig. 3.11). If the rock stays intact then one may find a complete rock sequence the wrong way up. If a fault forms, one may find a sequence of older rocks on top of younger ones. Nappes are common throughout the Alps.) The Iltay Nappe runs all the way from Deeside to Kintyre and probably right into Northern Ireland, and is a *recumbent fold* with the profile of a mushroom. The top and bottom halves join near the Highland

3.11 *Cross section through a nappe*

Boundary Fault where there is an anticline on its side that has since been turned upside down into a kind of syncline. As a result, large areas of rocks in the Southern Highlands are upside down because they comprised the lower half of this gigantic structure. Further to the north there is the *Balappel Nappe* that has an even more complex form!

The structures of rocks over much of the Highlands resemble those that can be made out of pastry or plasticine. Huge amounts of squeezing and crustal shortening must have been involved to make the rocks behave in this way – squeezing that came when the two continents to either side of the ocean came together again. Remnant pieces of ocean crust, of basalt pillow lavas and mantle rocks (altered peridotites), were emplaced along the Highland Boundary Fault and in the Southern Uplands. The rest of the ocean crust has obeyed the laws of plate tectonics and inconveniently disappeared. It has, however, in its absence been named *Iapetus* (after the mythical Greek father of Atlas whose name gave us Atlantic. It would perhaps have been more appropriate to name it after Atlas's mother, Clymene, who was herself a sea nymph).

The Iapetus Ocean has been traced between North and South Newfoundland, across Ireland and between Greenland and Norway, paralleling the position of the Atlantic, which only enters our tectonic story later (fig. 3.12). As the ocean closed the subduction zones that developed at either margin formed the andesite volcanoes of the Lake District and Wales as well as those now largely eroded off the Scottish mountains.

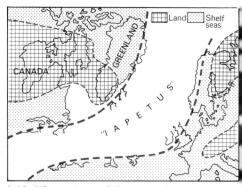

3.12 *The position of the pre-Caledonian 'Iapetus Ocean'*

Thick sedimentary sequences to the north continued growing out into the ocean as the mountains started rising behind them, right up to the last moment when the two sides came together at the end of the Silurian around 400 million years ago. It was in the lead-up to this collision that the major faults and folds of the Highlands developed. The Moine Thrust was the latest, and is therefore the best preserved of a series of thrust movements towards the end of the collision. Five major periods of folding have been unravelled for some areas, each one showing a different direction or a different kind of deformation.

To the south of the Highlands, the Southern Uplands is a tightly folded thick pile of sediments of Ordovician and Silurian Age, that kept forming in a trench that gradually moved south as mountains were piling up behind. The actual boundary between the two continents is well obscured; the two sides are now firmly interlocked. Hadrian's Wall (fig. 3.13) may co-incidentally mark the Northern boundary of both the former mid-European Plate and the Roman Empire.

3.13 *The possible former plate boundary above the lost Iapetus Ocean: Hadrian's Wall.*

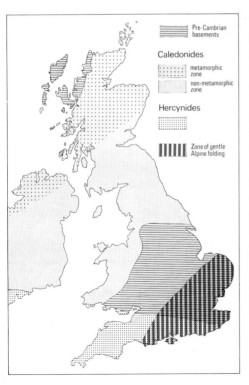

Pre-Cambrian basements

Caledonides

metamorphic zone

non-metamorphic zone

Hercynides

Zone of gentle Alpine folding

The whole confusion of events that took place in the destruction of the ocean and the construction of the *Caledonian* mountain range is termed an *orogeny*. Further south, in England and Wales, the deformation was much less severe, with both folding and faulting on a much smaller scale in Wales and the Lake District and across other parts of Britain not at present exposed (fig. 3.14). However, one feature of the orogeny to the south, the intrusion of granite magmas (see Chapter 4), was to have an important effect on the distribution of sediments ever since. These granites are lighter than the surrounding rocks and have stayed as higher ground while the land around has been sinking. To accommodate this, faults have developed around

3.14 *Deformational episodes in Britain* (left)

39

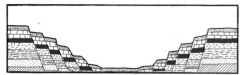

3.15 *Cross section through a rift valley*

them, as around the North Pennines, the Lake District and parts of North Wales.

After a period of relative calm, a new orogeny formed from the destruction of an ocean (or oceans) to the south, culminating at the end of the Carboniferous Period when the rocks of South West England, South Wales and South Ireland were folded and faulted along E-W axes, that contrast with the NE-SW axes of the Caledonian. A major thrust zone developed along the south Cornish coast, and can be seen at Start Point and the Lizard, with mantle and sedimentary rocks, of unknown origin, pushed on top of those of the Devonian. Throughout the rest of Britain, earlier faults and folds became reactivated. Faulting is particularly common in the British coal fields.

The most recent local orogeny was in the Alps, culminating only 10-20 million years ago. The long gentle folds that form the Chilterns, the North Downs and South Downs as well as the sharp folds running from Purbeck to the Isle of Wight, are 'the ripples of the storm of activity' taking place hundreds of kilometres further to the south east.

However, since the Permian and possibly even since the Devonian, Britain has been adjacent to the other end of the plate tectonic conveyor belt – that of new ocean formation. Cutting down the centre of the North Sea, buried beneath thousands of metres of sediments, is a major rift or *graben* (the German for rift) in which a central block has sunk leaving a buried steep sided valley, similar to the Rhine Graben, with which it interconnects (fig. 3.15). *Rift valleys* form where the crust is in tension. To the east and west of Britain there are rift zones and it is they, more or less, that have turned us into an island. Within the North Sea they stretch for 1200km (fig. 3.16).

The East African Rift, the Rhine Graben and the North Sea Grabens have all been the sites of alkali basaltic volcanics, indicating minor amounts of mantle melting below.

In the north North Sea the collapse of the crust had already begun in Devonian times, as basins, containing up to 10km of Devonian sediments, formed next to the rising mountains. For much of the area it is not known what lies below. A small graben of a sort already existed between the Lake District and the Isle of Man through Cheshire and along the Welsh border to Devon. At the end of the Carboniferous, at the same time as the mountain building to the south, crustal strains started to try to tear off Britain from Scandinavia and make an ocean where the North Sea now lies. Most recently, we can see the processes whereby a rift zone becomes widened into the beginnings of an ocean in the Red Sea, where ocean crust spreading only began a few million years ago. Why the first attempt in the North Sea failed, we don't know. Perhaps the forces resulting from the collision to the south dissipated too fast.

The next major attempt at splitting a new ocean came from the south west

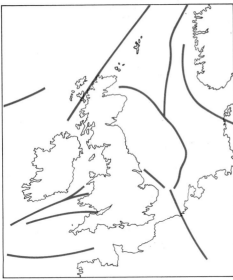

3.16 *The rifts around the British Isles*

towards the beginning of the Cretaceous, when the split between West Africa and America started moving north and succeeded in separating the Rockall Bank, that was then joined to Greenland and Canada, from Britain. The Rockall Bank has now been incorporated back into the British Isles through an Act of Parliament. A single, absurd rock sticks some 20m out of the ocean, washed over by the storm waves (fig. 3.17). Rockall is, in fact, separated from Britain by ocean crust and on geological grounds could belong to either Greenland or Canada. It became disowned by these Western continents

3.17 *Rockall – all that is exposed of a lost North Atlantic continent*

at the end of the Cretaceous and the beginning of the Tertiary, when the North Atlantic finally came into existence and ocean began to form on the far side of the Rockall Bank and between Greenland and Norway.

In the last fifty million years this small continent has sunk – a little too long ago to be the mythical Atlantis, but during the low sea levels of the last Ice Age – much more land would have been exposed.

The progress of this ocean split, as it started in the south, and gradually advanced north over a hundred million years, is one of enormous complexity. The ocean born from Iapetus almost chose so very many other paths. As the Atlantic continues widening at about 2cm each year we will never know how close we came in the Permian to leaving Europe behind and becoming the eastern edge of Greenland and North America (fig. 3.18).

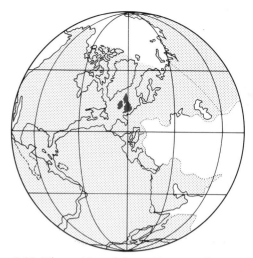

3.18 The position of the continents in the Permian before the opening of the Atlantic

GRANITE

4

The Intruders

For much of geology, the principals of *uniformitarianism* (that the present is the key to the past) have proved a useful technique for interpreting old rocks and old processes. There are, however a number of processes that we can never hope to observe – those that happen deep underground. If the magmas (considered in Chapter 1) fail to erupt, they may form intrusive or *plutonic* bodies of *igneous rock*. In general, while a basalt magma is likely to erupt, the more viscous granitic magma tends to remain underground. Granites are common in all orogenies.

In the 18th Century granites were the focus of a controversy, caused because no-one had ever seen one forming, between the neptunists who thought they crystallised from water, and the plutonists who thought granites had come direct from a hot melt. We now know that, as with so many controversies, the truth lies somewhere in the middle, though nearer to the plutonists. Granites seem to have crystallised from rather water-rich magmas.

Granite rocks are hard, so they are often used as kerb stones and the parts of buildings that need the strongest materials (fig. 4.1). The crystals of a granite are much larger than those of a

4.1 Quarrying granite for building stone, Cornwall (see also Colour 4)

The grain size of an igneous rock depends on how much supercooling exists within a magma as it crystallises. Supercooling is when a magma is below its solidification temperature, but has not started to crystallise because of the difficulty of making ordered crystal structures grow from nothing – termed nucleation. The greater the degree of supercooling, the more nucleii will form because the more unstable the magma actually is. The faster a magma cools the greater the chance of supercooling. Hence a magma erupted at the surface will cool fast, form many nucleii and be fine grained (fig. 4.2). The granite is coarse grained (fig. 4.4).

4.2 Volcanic: Basalt

4.3 Plutonic: Gabbro (Pyroxene (black) and Feldspar (white))

4.4 Plutonic:Granite (Quartz (translucent), Feldspar (white) and Biotite Mica)

volcanic rock because the magma cooled very slowly. The large body of magma underground holds its heat much better than a lava at the surface. The minerals inside a granite can be clearly seen and identified with the

naked eye. They are: quartz (which is generally translucent and colourless); one or two feldspars (one of which may be white, the other pink); and smaller amount of *mica*, which has a characteristic flakiness and glint. The mica is sometimes silvery white (called muscovite) and sometimes black (called biotite). The granite is rich in silicon and aluminium (as oxides) and is therefore very light coloured. The rarer rock, gabbro, that is the plutonic, coarse grained version of a basalt, contains pyroxene (dark grey-brown or black) and feldspar (white) with sometimes a little olivine (light green) (fig. 4.3). The combination of dark and light minerals makes a gabbro look speckled like a bird's egg. The ratio of the dark minerals (rich in iron and magnesium) to the light ones is a useful way of identifying these plutonic rocks (see fig. 1.5).

The plutonic rock made up entirely of the dark minerals, including olivine, is known as a peridotite. The mantle beneath the earth's crust is largely composed of peridotite. Although there is no volcanic equivalent of the peridotite it seems that early on in the history of the earth, higher temperature conditions allowed a peridotite magma to erupt. The melting point decreases as the proportion of light coloured minerals increases. Thus a magma of granitic composition can be a liquid at the lowest temperature of all.

If one heats up ordinary sediments and volcanic rocks (typical crustal materials) with some water, in a laboratory experiment, the first composition of liquid to form, at about 600°C, is a granite. When one heats up mantle peridotite rocks, under the conditions expected at depths in the earth, the first liquid to form in any appreciable quantities is a basalt magma. Thus granite is to the crust, as basalt is to the mantle. But as the underlying movements of plate tectonics only involve the mantle and not the rigid continental crust, the crustal materials only get hot enough to melt where rocks are squeezed and dragged down at subduction zones and in colliding continental plates. Andesites lie between basalts and granites in composition. They seem to form as a mixture of mantle and crustal melting processes under the subduction zone.

There is another way that granitic magmas can form; for if a magma of tholeiitic basalt crystallises very slowly, the high temperature minerals (pyroxene and olivine and the calcium rich feldspar) may sink to the bottom, taking with them most of the iron, magnesium and calcium and leaving behind a magma that is rich in silicon and aluminium of granitic composition. Around the remnant base of the Skye volcano there are numerous small granitic intrusions, some formed from the melting of the crustal rocks below by the heat given out by the basalt magma, and some from the last stages of the crystallising basalt magma itself. However such granites associated with basalt volcanoes are much less important than those that form in orogeny.

Our most recent orogenic granites are those of the Hercynian earth movements at the end of the Carboniferous, that were centred further to the south in mainland Europe, but that also reached into South Western Britain. In Cornwall and Devon the amount of uplift and erosion since that time has been just sufficient to reveal the tops of a series of large granite intrusions, that were originally buried under a few kilometres of overlying sediments.

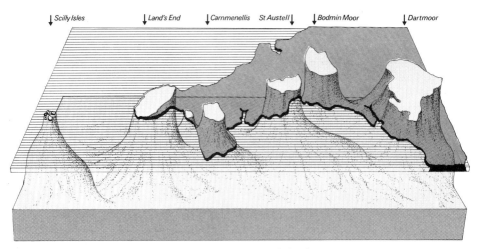

4.5 *The exposed and hidden batholith of South West England*

There are five main ones, trending in a row **WSW** from Exeter to Land's End, with the Isles of Scilly forming an offshore extension of the same series (fig. 4.5). They were all intruded at the same time – 290 million years ago. Through mapping the pull of gravity over the whole area it has been found that the low values found along the whole line can best be accounted for by the presence of a large underground granite intrusion that connects them all. (Granite is lighter than the average crustal material and therefore reduces the pull of gravity above). The granites we see are just the higher domes, or *cupolas*, of a massive intrusion, or *batholith*, about 12km underground (fig. 4.7).

The granites are harder than the sedimentary rocks into which they were intruded, and have all formed upland areas, or plateaus, or in the Scillies, a group of islands. The granite outcrops generally contain the characteristic horizontal cracks or joints that give rise to the famous *Tors* (fig. 4.6). These have formed through a combination of chemical and physical weathering that takes advantage of easy access to the rock along these planes of weakness. The apparently horizontal cracks, formed as the granite solidified, are parallel to the shape of the intrusion. Where granite intrusions are better exposed, as in California or Africa, the granites have eroded (*exfoliated*) along these surfaces, making perfect dome shaped mountains (fig. 4.7). In Devon and Cornwall we are merely seeing the top and therefore the horizontal parts of these domes.

The water in the original granite magma has not yet entered our story, but as the granite cooled, the water could find no home within the minerals and so became channelled out through the high level domes of the intrusion, causing considerable alteration and havoc. The massive feldspars (up to 20cm long) (fig. 4.8) that are common in several of these

4.6 A Dartmoor granite tor – Haytor Rocks

4.8 Feldspar phenocrysts in granite

intrusions kept growing across the other minerals after crystallisation because of the extra potassium supplied by this fluid. Towards the top of the granites, where the hot and reactive water or steam would naturally tend to accumulate, the granite was broken down. The feldspars and micas of the granite were altered to a very fine grained clay mineral called kaolinite

4.7 Eroded top of a granite batholith; weathering along curved cooling surfaces – Half Dome, Yosemite, California. For scale, see centre background figure 9.8.

4.9 *Thin sections of granite altering to Kaolinite (a) granite, (b) kaolinite and quartz*

and only the quartz was left intact (fig. 4.9). The conditions were rather like some very high temperature and very rapid form of weathering. These large deposits of *hydrothermally*, (or hot water) altered granite lie on top of the intrusions and are now extensively quarried for *kaolinites*. Kaolinite is the perfect clean white China Clay mineral that can be moulded and then fired, to drive off the water, and make some new minerals (aluminium silicates) of a man made rock we call 'china' or 'porcelain'. It is also extensively used in paper, plastics and many other industries. The lunar landscape seen around the China Clay pits results from the large piles of quartz waste that is left behind after the clay is extracted with jets of high pressure water (fig. 4.10).

Other parts of the granite intrusions have sometimes been affected by localised concentrations of boron and fluorine; elements that were also left behind when the magma crystallised. These have formed spectacular radiating clusters of the mineral tourmaline, of aesthetic rather than economic significance (fig. 4.11).

Some of the hot water from the granite started on a different journey.

4.11 Radiating needles of tourmaline in altered granite

4.10 China Clay quarry and 'Cornish Alps' (spoil tips), near St Austell

Enriched with elements left over after the granite crystallised, as well as some collected from the breakdown of micas, these hot fluids set off along cracks in the surrounding rocks. These cracks, probably made by the intruding granite, were straight for long distances rather like very thin dykes. As the fluids travelled, the temperature decreased, also the pressure, for they were moving towards the earth's surface, to emerge as hot springs.

As the cooling water could no longer carry all the dissolved elements, they were deposited. First of all the tin and the tungsten, then the arsenic, the copper, the uranium, nickel and cobalt and finally the silver, lead and zinc were left behind in ore minerals, as sulphides or oxides according to the conditions and the stabilities of the various forms. Quartz, feldspar,

tourmaline and calcite minerals were also formed, again depending on the changing conditions along the passage of the fluid. These infilled cracks around the granite intrusions are known as *lodes*. Many of them have been mined in order to obtain the elements from these minerals, that the fluids carefully extracted and concentrated from the original granite intrusion. One of the copper mines, the Devon Great Consols, was once the biggest in the world!

The pattern that these lodes make is illustrated for one of the granites in fig. 4.12. Although they can be followed down to deep levels, these cracks are somewhat contrary. Tin miners have to have a good instinctive feel for geology, for many of the cracks were faults before the mineralisation took place, and different veins at different heights may be enriched in a particular mineral. The reasons why such rare elements as tin or tungsten should be carried at low concentrations within fluids, to be deposited in very high concentrations, are unclear. However such mineral veins are common around granite intrusions that have later been subject to some hydrothermal attack. Similar veins radiate from the Skiddaw granite in the Lake District. Many important deposits of copper minerals, throughout the world, seem to have formed from fluids and gases associated with andesite magmas and this is true of both the Lake District and North Wales, where copper mining was once a major industry.

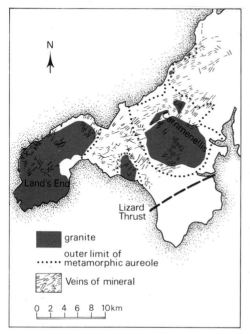

4.12 *Map of mineral veins (lodes around a granite intrusion)*

4.13 *British Granites and Gabbros with areas considered for potential geothermal energy abstraction*

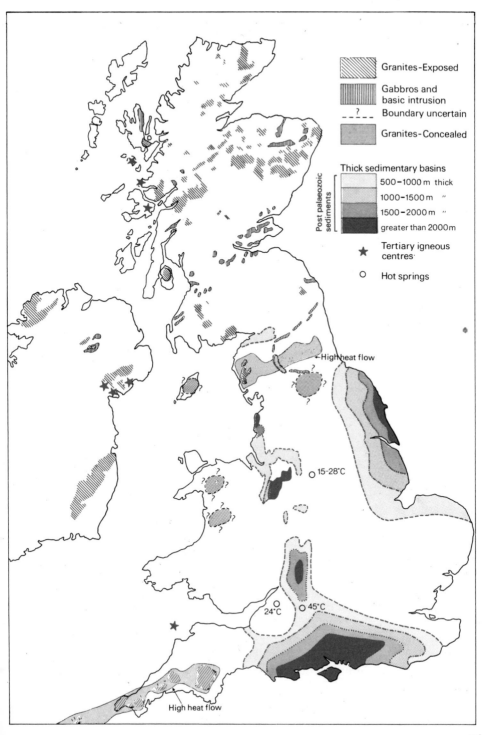

Granites-Exposed

Gabbros and
basic intrusion

? Boundary uncertain

Granites-Concealed

Thick sedimentary basins
500–1000 m thick
1000–1500 m "
1500–2000 m "
greater than 2000m

Post palaeozoic sediments

★ Tertiary igneous centres·

○ Hot springs

←High heat flow

15-28°C

24°C

45°C

High heat flow

4.14 Vein of cassiterite ore in Cornish tin mine

There are more than fifty separate granite intrusions in Britain (fig. 4.13). Almost all the major ones are associated with the Caledonian orogeny, and about thirty of them are in Scotland, though the largest is the Leinster granite of SW Ireland. One of the prettiest of all building stones comes from the Shap Granite (Colour 3) on the edge of Cumbria. A low gravitational pull extends from the Shap Granite across to the east to connect up with the Weardale Granite of the same age (410 million years old) that is entirely buried beneath Carboniferous limestones and only ever seen in boreholes. This granite appears to have originated in upper mantle rocks from some kind of process involving an ancestor as a basaltic or andesitic magma.

Another buried granite, a little to the south, the Wensleydale Granite, has the same age but appears to have formed from crustal melting. These granites have provided support for the theory mentioned in Chapter 3; that the boundary of the plates and the remains of the Iapetus Ocean lie along the eastward extension of the Solway Firth.

Much of the uplift of crustal blocks throughout Britain's history can be related to the granites. The fairly recent doming of the Lake District, for instance, can be attributed to the buoyancy of the massive granite intrusion now known to lie beneath it.

The giant granite batholiths, found above former subduction zones in the Andes and in North West America, are missing from this country, chiefly

4.15 Migmatite – mixture of light coloured granite and dark layers of unmelted rock

because the uplift and erosion has not been sufficient to reveal them, even though they may exist at depth, as in Devon and Cornwall. The largest of all granites may be a hundred kilometres in length and of enormous volume. What lies beneath them?

Fortunately we can piece together a picture of what lies beneath from rocks that have come from different depths and are now found in different parts of the country. In parts of the Highlands there are large areas of rocks known as *migmatites*, or sometimes as the Older Granites. A migmatite is a striped rock consisting of granite veins thoroughly mixed up with material rich in darker minerals (fig. 4.15). They are rocks that had started to melt. Originally homogeneous sediments have been heated up under pressure to 700 or

800°C. A granitic magma was sweated out to form veins while the dark minerals failed to melt and were left behind. Here we see a granite in formation. Although some of the magma may have migrated out of the area into the rocks above, and been eroded away since, fortunately some stayed behind. These Scottish migmatites formed very deep down – perhaps 15km underground, about the same depth as the base of the hidden Cornish batholith.

But in asking 'why?' we can go back a stage further. Why did these crustal rocks melt? In the same area as these migmatites there are large bodies of gabbro of the same age. The gabbro (or basalt) magma was probably about 1000°C – hot enough to cause the melting of nearby rocks. The gabbro magma came from deeper melting – probably connected with convection or disturbance within the Upper Mantle. The only gabbro in England is at Carrock Fell near the Skiddaw Granite, but unfortunately, the link between these two rocks is not so clear cut.

In terms of searching for possible sites for taking heat from the ground to make *geothermal power*, the high heat flows and granite intrusions of Britain are a thing of the long distant past. Though the heat flow from the Cornish granites is still anomalously high, the most suitable locations for finding hot rocks near the surface are where a thick layer of insulating clay has prevented the normal heat of the earth escaping, instead encouraging it to build up in the rocks immediately beneath. In the Paris basin, heat is already extracted from such a sedimentary rock trap and it is possible that similar British rock sequences, such as exist within the Hampshire basin, could also provide geothermal energy (fig. 4.13).

Granites haven't quite lost all their mysteries. We can still never hope to see them forming, although there are places in the Andes and in Japan where it seems as if deep down, a big granite intrusion is developing today. Still, the study of granites has come a long way in a fairly short time. Only thirty years ago, an extraordinary fluid termed 'ichor' (the blood of the Greek Gods) was thought to be running through veins in the bowels of the earth, busy turning the sediments to granite!

MINERALS

5

All that glisters...

From the earth, man has taken the materials that have allowed him to make his own environment – the stones for building, the metals and the fuels. Other rocks have always been collected for their beauty. Soon the beauty and scarcity of certain rocks became mixed together in the idea of value. Gold, diamonds and other precious jewels have now become ways of storing money itself. Fortunately there are many attractive stones and minerals that are not rare enough to have gained a price on their heads! In this Chapter we will have to look at what rocks are made of and see the kind of conditions required to produce good rock and mineral specimens.

All rocks, except volcanic glasses, are made of mineral grains, and each grain is a *crystal* (fig. 5.1). Glasses are unstable liquids that failed to crystallise. Some rocks, such as a sandstone (quartz) or a limestone (calcite) are made up of just one kind of mineral. Others, for instance those that, like granite, crystallised from a magma, may be made from three or four different mineral types. Unlike the atomic disorder that exists in a glass, the atoms of a crystal are stacked against one another in a regular ordered arrangement. A tiny unit of the crystal pattern, containing only a few atoms, is exactly repeated millions and millions of times to make up the smallest crystal.

As a rule, at low temperatures, stacking atoms in an ordered fashion can take up less space than a disordered one (fig. 5.2). Of course, actual crystal

5.1 *Crystallites forming in an unstable glass*

5.2 *Order and Disorder – saving space through regular stacking*

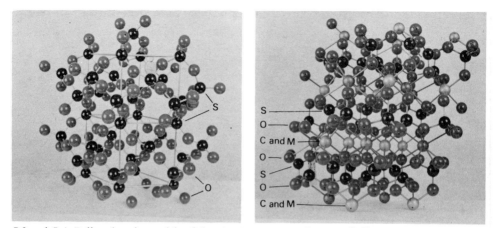

5.3 and 5.4 Ball and spoke models of the atomic structures of quartz (left) and pyroxene (right). Symbols for the atoms shown are (O = Oxygen; S = Silicon; C and M = Calcium and Magnesium)

structures are much more complex than our analogy. Only the very simplest of elemental minerals contain just one kind of atom. Most crystals are built from two or three different atoms, some such as tourmaline from as many as eight different atoms, each with their own special site within the crystal structure. The atoms are held together by the interchange of certain electrons that orbit around their central nucleus, forming a chemical bond.

The minerals that make up the rocks are nearly all silicates (made out of silicon and oxygen). The simplest of these is *quartz* in which the silicon and oxygen atoms are bonded together as triangular pyramids or tetrahedra with one silicon atom in the middle surrounded by four oxygens at the corners (fig. 5.3). These are then joined to one another to make a framework that is the crystal structure. *Feldspars* have a rather more complex kind of framework with some aluminium and calcium, sodium and potassium atoms present. *Pyroxenes* have chains of silicon and oxygen atoms joined together by iron, magnesium and calcium atoms. (fig. 5.4).

There are around four thousand naturally occurring minerals found on earth each with its own variation to the problems of stacking certain kinds of atoms against one another. Each mineral structure represents a solution to this problem for a very limited range of chemical compositions. The chemistry of a normal rock therefore has to be accounted for by a mixture of minerals.

Many properties of crystals are related to their internal structure and symmetry. The most familiar is the external form. Although all the minerals that go to make up the rocks are crystals, rich in internal symmetry, one would not recognise this from their shapes. In most rocks there is no free space so the crystals are hard pressed up against one another. Their outside surfaces are thus a result of competition (see fig. 4.9A). Crystals can only grow to their ideal with their shapes imposed from the inside if the competition from the outside forces or

Order within a mineral structure means *symmetry*. Symmetry is not just the way the small section of the pattern or the arrangement of the atoms is repeated (like tiles on a floor) but is also the way that the atoms inside that unit relate to each other (fig. 5.5). To show why crystals have symmetry within their

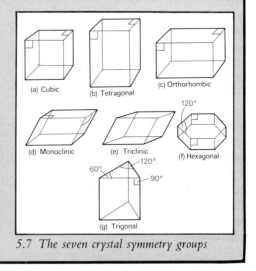

5.6 *A harmonic of oscillation for a three fold axis of symmetry – three atoms bonded to a central one*

5.5 *Symmetry in Moorish wall tiles*

order and particularly why (in some cases) the symmetry of crystal structures tends to increase with temperature, it is necessary to remember that all solids are vibrating and that as all the atoms 'hum', symmetry is a way of adding extra harmonics to the vibrations. For instance, if all the atoms around a point are repeated exactly three times then a three-fold wave form or vibrational harmonic can be set up around that point (fig. 5.6). The more symmetry, the more harmonics of vibration and the more energy can be stacked in a crystal. The kinds of symmetry that can exist in crystals can be considered separate from the actual atoms of crystal structures themselves. There are seven symmetry groups; their outside shapes are represented in fig. 5.7. Within these shapes there are 32 ways of arranging the various kinds of symmetry units

whereby one piece of a structure repeats somewhere else, and there are a total of 230 ways of arranging the actual symmetry of the space inside the crystal. To understand all the workings and fascination of symmetry really requires a book all of its own. It is remarkable to think that in all the chaos of the rocks, at the smallest level, there is perfect order.

5.7 *The seven crystal symmetry groups*

from other crystals is very weak. This is only possible when the crystal grows into an open space or into surrounding material that is much softer. Well

formed crystals of *garnet* may form in some metamorphic rocks, or *pyrite* in clay because they are much harder than the surrounding material (fig. 5.8 and

5.8 *Cubic crystals of pyrite in schist*

5.9 *Varieties of garnet crystals in schist*

5.9). All things being equal, when there is little opposition the atoms will add themselves to some planes of the growing crystal better than to others. The relation of these outer planes to one another will be controlled by the same symmetry that relates the atoms of the smallest unit of the structure.

The best and prettiest crystals have grown naturally in open spaces in the rocks. Such open spaces are rather uncommon but can be formed, perhaps where a fault movement has shifted from one plane to another (fig. 5.10); or as gas bubbles inside some volcanic lavas; or as cavities left by solution in a rock such as a limestone. In the footsteps of the miners, who followed the cracks and faults of the rock to trace mineral veins of copper, lead, zinc, tin and tungsten, went the early mineralogists to collect the best crystal specimens. These crystals were often not of the minerals being mined, and so would be discarded.

As long as the crystals that grow from the ore fluid in a fault cavity do not meet and destroy each other's delicate faces, such cavities will contain the kinds of crystal tableaux found in museums (Colour 4). The size, colour and artistic beauty of the display is left up to the discretion of the ore forming solutions! Unfortunately these ore veins were mined as deep and as far as was possible and to find similar museum specimens today would require re-opening the mines.

A more likely place for finding good crystals is in volcanic gas bubbles (fig. 5.11). While the lava is still warm, hot solutions may pass through the rock causing new crystals to form in the spaces. These spheres of inward pointing crystals are known as *geodes*. As the lavas are often quarried for road stone, it is still possible to find good geodes exposed in quarry walls. It is an extraordinary thing to crack open a dull brown sphere of rock to find the inside is a glittering mass of crystals.

The most common minerals to grow as well formed crystals are quartz and calcite. Calcite crystals are most common in cavities within limestone, for limestone is itself made out of fine grained calcite. (*Calcite* is calcium carbonate.) The normal crystals are small, three sided spikes (fig. 5.12), but one may also find 'rhombs' of

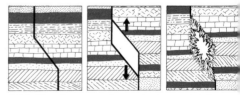

5.10 *Development of crystals in gap left by movement along a shifted fault plane*

5.11 Unbroken and broken geodes – rich in internal quartz crystals (left)

5.12 Calcite crystal form – dog's tooth spar (below)

5.13 Double refraction on the cut faces of Iceland Spar, a form of Calcite (bottom of page)

calcite that have the curious property of making two images of anything seen through them (fig. 5.13). This is because the light is bent, as it is when viewing an object that is underwater, but in calcite the crystal structure has determined that it is bent much more in one direction of the crystal than in the other, splitting the image into two.

Quartz crystals form in the shape of a hexagonal prism capped with a hexagonal pyramid (fig. 5.14). The

5.14 *Quartz crystal form Hexagonal prism tipped with hexagonal pyramids.*

internal crystal symmetry is actually 'trigonal' (like calcite) but quartz is very close to being hexagonal. Quartz crystals commonly contain small amounts of impurities that affect the colour, and perfectly clear quartz, or 'rock crystal', is fairly rare. The commonest colour is white and cloudy, translucent rather than transparent with many flaws and imperfections. *Amethyst*, the purple-lilac form of quartz contains impurities of iron oxide; rose quartz contains a trace amount of manganese oxide and smokey quartz with a grey-brown gloom is from radiation damage due to impurities of radioactive elements (Colour 6 and 16).

Most of the really precious stones cannot be found in the British rocks. Diamonds, emeralds, jade, sapphires and rubies (the last two are both impure forms of corundum) can only be found in museums and jewellers' shops where the original crystal will be cut with some new faces to allow it to focus the light and acquire 'brilliance'. However, some of the so-called 'semi precious' minerals can be found: garnet and zircon in the Highlands; topaz in parts of Scotland as well as parts of Cornwall, where one can also discover crystals of tourmaline. *Fluorite* cubes are common in limestones of the Carboniferous age where they accompany veins of lead and zinc minerals (Colour 8).

Fluorite (calcium fluoride) was mined in its own right at Castleton in Derbyshire, where thick veins of striped purple-blue-yellow and colourless fluorite known as '*blue john*' (from the French: bleu-jaune) were carved into vases or inlaid in tables (Colour 7). Elsewhere it was discarded on mine spoil tips to be picked up by amateur mineralogists, and more recently by industry that has found a use for it in metallurgical processes.

Under some conditions, crystals may grow to an enormous size. Such conditions are rather like those required to grow a prize-winning vegetable marrow – plenty of nutrients and time and nothing to get in the way. Such conditions are found in the geological world in *pegmatites*: thick dyke-like veins that may channel off a water and silicate rich fluid from the end stages of the crystallisation of a granite. Being much less viscous than a granite, the crystals that nucleate at the margins can continue growing into the body of the fluid at great speed, by geological standards. A quartz crystal weighing 13 tons has been recorded in Siberia, a mica crystal weighing 7 tons from Ontario, and a crystal of the mineral beryl (rich in the element beryllium) weighing 40 tons from Malagasy (formerly Madagascar), (fig. 5.15). All these were formed in pegmatites. Such crystals may reach several metres in length!

Some kinds of meteorites, the nickel-irons, may be made up of a single crystal (but without straight sided faces). On a more mundane level it is even possible for a freezing lake to nucleate a crystal of ice to one side that may grow to cover an enormous area, sometimes the whole lake!

However, it is possible to find pleasing shapes and colours within rocks and minerals without them being quite so exotic. Some of the ore minerals are often very attractive: *galena* (lead sulphide) that forms highly reflective silvery cubes, pyrite (iron sulphide) that shines like gold and is also known as 'fool's gold'; or the duller creamier, more orange colour of *chalcopyrite* (copper-iron sulphide) are some of the most common. Alteration products of the copper minerals produce a rainbow of colours; green (as in *malachite*), blue (as in *azurite*) or

5.15 *Giant smoky quartz crystal*

red as in pure or native copper, *Magnetite* (iron oxide) occurs as steel grey cubes or granules in a number of rocks (Colour 9.14).

Some of the most prized rocks for the collector are those which occur as nodules or spheres. When split open growth. *Nodules* include: agates, that formed through the slow growth of formed through the slow growth of different coloured bands of quartz inside geodes (Colour 15); radiating crystals of the mineral *hematite* (iron oxide) that grow out from a common centre to form globular masses termed kidney ore; *marcasite* that forms radiating aggregates of crystals of iron sulphide in the chalk, or *septarian nodules*, filled with shrinkage cracks, from the Oxford Clay.

Even the humblest pebble beach is flattered by the sea, that leaves a thin film of water over the surfaces of the stone to hide irregularities and show off the colours below. Pebbles can be given a similar shine by polishing them, so that the stone is smooth (Colour 17). The simplest way of polishing stones is to rub them against one another, in a rock tumbler, in much the same way that the rocks are smoothed as they collide with one another on a beach.

For those who study geology the beauty of a rock goes far beneath the skin. The prettiest patterns and the most fascinating and striking colour schemes can be found inside a rock. To study the minerals that go to form the rocks it is necessary to slice a 'thin section' only 0.03mm in thickness hardly thicker than a sheet of paper. The minerals, that from the outside seem dull and black, may take on new life when the light is allowed to pass through them (Colour 18). Like some abstract stained glass window, the brilliant colours not only please the eye but for the trained petrologist can help tell a story of how the rock was originally formed. And few stories are illuminated with such dazzling visual effects.

PRE-CAMBRIAN

6

Back to base

The first attempts to make some sense out of the rocks that outcrop across Britain became the science we now know as *stratigraphy*. Different amounts of uplift and erosion as well as different areas of original rock formation have presented Britain with a wide variety of rock types, building the scenery at the surface. Stratigraphy used the horizontal layers that form within the sediments and followed them for as far as possible. If there were gaps in the rock outcrop, then it was possible to *correlate* rocks from one place to another by looking at the fossils within them. Most organisms, throughout geological history, have been busy evolving, and the changes in their fossils have allowed rocks in one continent to be correlated or matched with rocks from another.

The most important of the ideas of stratigraphy is the simplest; that of *superposition*. The sediment on top is going to be younger than the material underneath.

The series of sedimentary rocks was divided up into episodes of formation.

6.1 Pre-Cambrian outcrops of Britain

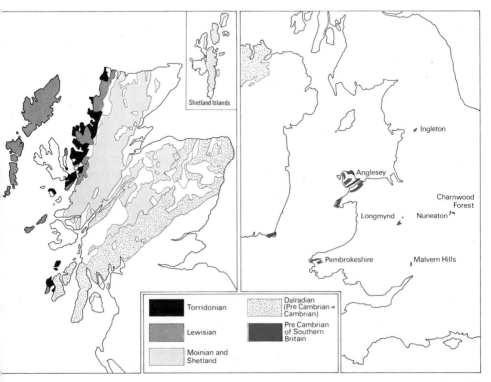

Shetland Islands

Ingleton

Anglesey

Charnwood Forest

Longmynd

Nuneaton

Pembrokeshire

Malvern Hills

■ Torridonian

▨ Lewisian

▦ Moinian and Shetland

▨ Dalradian (Pre Cambrian → Cambrian)

▦ Pre Cambrian of Southern Britain

These are the geological periods (see back cover). As many of them were defined from the British rocks, the boundaries were set where gaps occur in the sediments of Britain. It soon became apparent that although this made it easy to classify British rocks, it made it very hard in another country where breaks or gaps in the sediments lay at different times.

Thus a geological time scale was measured in rocks. The geological periods were the equivalent of time-units; the hours or years. Naturally, many people were interested in attempting to find the 'exchange rate' for converting this 'rock time' into real time. In 1897 the physicist Kelvin worked out that if the earth had started very hot and had cooled down without any heat sources of its own, then it could only be between 20-40 million years old. This was not nearly enough to allow evolution to have turned the single cell into a man, nor for the enormous thicknesses of rocks to have been deposited.

The new discovery of radioactivity solved (and started) many problems. If the earth contained radioactive elements (which it did) then the heat they produced would keep the earth hot for billions of years. Through studying the rate at which certain radioactive atoms decayed it was possible, for the first time, to put a

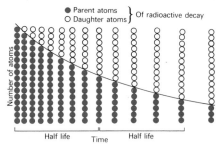

6.2 Radioactive decay – half life of a radioactive isotope

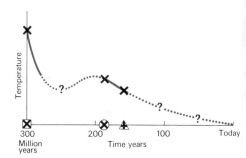

6.3 Different age measurements obtained from the same rock indicating a period of metamorphism since formation

⊠ Age from Rubidium 87/Strontium 87 for rock
⊗ Age from Rubidium 87/Strontium 87 for mica
▲ Age from Potassium 40/Argon 40 for mica

Continued on page 66.

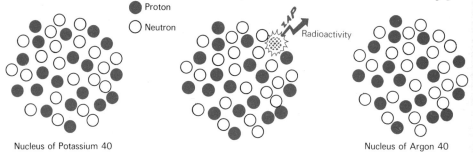

6.4 Radioactive decay of Potassium 40 to Argon 40

Certain of the natural elements have more than one form or *isotope*. (An isotope of an element has the same number of protons and electrons, but a different number of neutrons.) Some of the isotopes of the natural elements are unstable and will alter to other atoms with different nucleii and lower atomic weight, through the emission of some energy. This breakdown is called *decay* and the energy is known as *radioactivity* (fig. 6.4). The rate of decay is very variable from one kind of atom to another. If we measure how much radioactivity is given out over a short period of time we can find out what proportion of the atoms present have decayed and then how long it would take before there were only half as many of the initial sample of atoms remaining. This period of time is called the 'half-life' (fig. 6.2).

For an isotope to be useful for measuring the age of something it should be: a) abundant; b) have a half life of the same general order of age as the age of the material; and c) should have easily recognisable *decay products* – the atoms formed after the decay. The most important radioactive time keepers for the rocks have been: Uranium 238 (a Uranium isotope with an atomic weight of 238) that has a half-life of 4,498 million years, and decays to Lead 206; Uranium 235 that has a half-life of 713 million years and decays to Lead 207; Rubidium 87 that has a half-life of 50,000 million years and decays to Strontium 87; and Potassium 40 with a half-life of 11,850 million years that decays to Argon 40. By measuring the amount of decay product in relation to the amount of the unstable parent isotope it is possible to find the *age* of a rock.

During the processes of weathering, or the crystallisation of a rock from a magma, the parent and daughter atoms of these radioactive breakdowns may become well separated. Thus the *age* of a rock can measure its formation. However, if the rock has been heated up since it was formed, then the daughter atoms may move out of the minerals or even out of the rock (for instance the argon that forms from potassium is an inert gas that will tend to escape). Thus the *age* may record some event in the rock's past other than its formation. By using different isotopes in individual minerals or the whole rock it is possible to map the history of the rock as it passed through a period of being heated (fig. 6.3).

In all cases, very small samples of rocks or minerals are used in order to find a reliable estimate of age, and so it is necessary to measure the amounts of these atoms to an incredible accuracy. This is possible on a *mass spectrometer*; a machine that identifies the isotopes by turning them into charged ions. These are sent hurtling through a magnetic field, that separates them, and sends specific isotopes to the target, where the arriving atoms are counted (fig. 6.5).

6.5 *Technology is now important in many areas of geology in which a very accurate analysis is required.*

6.6 *Outcrop of pre-Cambrian rocks at Charnwood Forest, Leicestershire*

6.7 *The pre-Cambrian fossil Charnia Masoni from Charnwood Forest*

date on when a rock was formed. Because of the radiometric time-scale, we are no longer so dependent on the 19th Century logic of stratigraphy. We can find an order within parts of geology that are too complicated to be looked at from superposition and that are barren of the all important fossil time markers. Such is the *pre-*

Cambrian. It is only the name that is an unfortunate hang-over from the age in which such rocks were declared to be from the unresolvable chaos that existed towards the beginning of the world. The naming might have been different if the rocks had been defined from somewhere outside Britain, where pre-Cambrian formations can be thick layered sedimentary rocks containing fossils. More ridiculously, the geological periods each last only 40-80 million years until one goes back to the pre-Cambrian, that contains rocks spanning more than 3,000 million years!

South of the Caledonian mountains pre-Cambrian rocks outcrop in a few isolated places in the English Midlands, and both North and South Wales (fig. 6.1). The most easterly outcrop is at Charnwood Forest in Leicestershire (fig. 6.6). It is a buried mountain that protruded above a desert in Triassic times, and is just, through erosion and quarrying, beginning to re-emerge. The rocks consist of volcanic ashes (now made into a hard rock known as 'tuff'), overlaid by a conglomerate of largely volcanic fragments, followed by slates and grits in which a lone pre-Cambrian fossil, named Charnia Masoni (fig. 6.7) was found by a schoolboy some 20 years ago. The whole sequence has been folded and intruded with magmas similar to, but less quartz-rich than granites, known as syenites, that give ages of 680 million years.

Further to the west, pre-Cambrian rocks form the hills of the Longmynd in Shropshire, with 5,000m of mostly shales, sandstones and conglomerates that were preceded by acid volcanics. The sequence has been curled into a huge fold cross cut by dykes, all before

6.8 The outcrop of hard crystalline rocks making up the Malvern Hills, Worcestershire

Cambrian sediments were deposited on top. Further to the south, along the Wales–England borderland, there is the pre-Cambrian of the Malvern Hills (fig. 6.8). These include plutonic intrusions of gabbro and granite and metamorphic gneisses (rocks that have suffered a period of being buried a long way down in the earth's crust). This section of rocks must have suffered considerable uplift and erosion before the formation of the ordinary Cambrian sediments that lie on top.

In North Pembrokeshire, there are pre-Cambrian rhyolite volcanics that match with similar rocks above the largest area of the Southern British pre-Cambrian, at Anglesey. This comprises an incredible 10,000m of volcanic and sedimentary rocks. The lower part of this sequence contains *greywackes* mixed in with intrusions of gabbro and peridotite (that have been altered to the mineral serpentine) and an upper

portion of limestones, sandstones and other sediments, with a massive zone of large scale *shear*, a giant fault zone as opposed to a fault plane, cutting through it all. Many of the rocks have been highly folded (fig. 6.9) and some have suffered various degrees of change through temperature and pressure, or just through pressure in the shear zones.

The whole complex of these rocks is interpreted as having been involved in a subduction zone, that was in operation around the time of the pre-Cambrian/Cambrian boundary. Large scale mountain-building activity around this time in England and Wales would explain the sharp break that allowed the early geologists to define the boundary. But how to fit in the other lonely Welsh and English outcrops?

Since then, most of the area has sunk at successive periods building up large thicknesses of sediments that have

67

6.9 Mona complex of pre-Cambrian in Anglesey

'drowned' the pre-Cambrian rocks and allowed only these few small 'islands' to protrude through. Relating these islands is not only difficult because they are well separated, but also because the rocks appear to change radically over small distances. The volcanics of North and South Wales, as well as some of those further to the east, may be from occasional volcanic activity above the subduction zone. On the edges of these islands, the pre-Cambrian rocks dive down beneath the later simpler cover formations. Beneath the thick basin of sediments across North West Wales it dives so deep as effectively to vanish, but further to the east, the pre-Cambrian can be found between outcrops in boreholes at quite shallow depths. For much of the central eastern area of England the hard rock *basement*

is only a 'field's length' beneath the soft rock cover. For much of the northern part of the area, it is the pre-Cambrian that is this basement. Further to the south there are highly folded rocks of slightly later ages (fig. 6.10).

Towards the North of England, the sediments thicken up again on top of the basement and the pre-Cambrian is missing except for a small area of outcrop of deformed greywackes and volcanic debris at Ingleton, that lies on a continuation of the S.E. Ireland–Anglesey pre-Cambrian ridge.

Beyond there is nothing, until one reaches the other plate to the Northern side of the Caledonian ocean join (see fig. 6.1) where the whole nature of the *continent* and the *basement* is much more straightforward. To the N.W. of the mountain belt that now forms the

Highlands, there is a fragment of typical thick continental shield crust, that was wrenched off Greenland at the recent opening of the North Atlantic. Evidently the glue used to stick the Caledonian mountain belt together, on the destruction of the Iapetus ocean, proved stronger than the actual fabric of the continental crust. Towards the end of the pre-Cambrian Period, an enormous thickness of sediments was deposited on this continent–ocean margin, with the *sedimentary wedge* advancing further from the shield area, until subduction, ocean closure and collision turned this whole mass of sediments into mountains. The sediments change from 7,000m of 'Torridonian' river deposited sandstones, mudstones and conglomerates lying on collapsed continental shield to the west, to a similar thickness of '*Moine*' sediments, deposited on the edge of the ocean to the east.

Many of the older rocks of Britain, like the Moine, but not the *Torridonian*, have suffered a period of *metamorphism* since their formation. Much of the sediment involved in the Highland Caledonian orogeny has suffered metamorphism. The degree of metamorphism, or the *grade*, is dependent on how much uplift and erosion there has been since that time. In place of mapping the stratigraphic structure of the region one can map the boundaries of increasing metamorphic grades. These grades can be established by studying the new minerals that are formed within the rocks. Certain minerals may only become stable at increased temperatures and pressures and the appearance of these minerals within the rocks can be used to map metamorphic *isograds* (rocks of the same metamorphic grade). The pattern of these is much simpler than the actual rock structure because the metamorphism took place after the worst of the rock deformation was over (fig. 6.14).

On top of the Moine sediments, but also more to the south and east of

Continued on page 73.

6:10 *Cross section through Southern Britain*

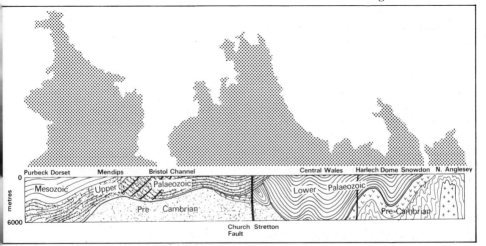

Metamorphism means a change or a transformation. Most metamorphism takes place because rocks become sufficiently deeply buried to reach high pressures and temperatures. The most usual place for such thicknesses of sediments to form and for *crustal downwarping* to take place is at the destructive plate margin; associated with subduction. High heat flow, in combination with the considerable up and down rock movements within the squeezing of orogeny and tectonics, means that sediments or volcanics can get dragged down and most important, re-uplifted, to reappear at the surface.

The effect of pressure and temperature upon rocks is to change first the shapes of mineral grains, and secondly the actual minerals themselves (fig. 6.11). Beginning with a mud, the transformation to a fossil mud or mudstone takes place through compaction and loss of water with a localised increase in grain size and rock strength. Such changes can take place purely from the weight of a relatively small amount of overlying sediments (about 1,000m). Further compaction may make a stronger harder rock. The onset

6.11 Metamorphic rock series (a) clay, (b) shale, (c) slate, (d) schist, (e) gneiss

(a)

(b)

(c)

(d)

(e)

of metamorphism is defined as being where there is considerable recrystallisation of the minerals.

Within orogeny there is always deformation, rock folding and faulting, accompanying the downwarping of sedimentary piles of rocks into metamorphic conditions. As the rocks are beginning to recrystallise they are being squeezed; in fact this squeezing will speed up the changes. The new mineral grains, that develop in mudstones at low temperatures, have sheet shaped crystals (like clay minerals and micas) and these all grow as flat crystals at right angles to the squeezing. The rock turns from a shale into a slate, for all these parallel sheet shaped crystals allow the rock to be broken along perfectly flat surfaces. The slates of North Wales and the Lake District are mudstones or volcanic ash deposits that have suffered mild metamorphism. These perfectly flat surfaces of slates make them ideal for the construction of billiard tables. The way in which the rock can be so neatly split allows it to be used in roof tiles. The ability of a rock to split into layers is known as *cleavage*.

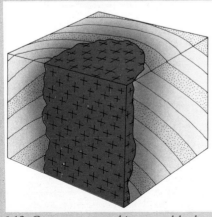

6.12 *Contact metamorphism caused by heat from an igneous intrusion*

At deeper levels of burial and metamorphism the crystals inside the rock become still bigger and more easily visible. As they increase in size the cleavage becomes rougher and the sheet shaped mineral grains become large enough to give the newly broken surface of the rock a glitter. The rock is now a *schist*. Further metamorphism will separate pods of light and dark minerals, which may be made up of crystals up to a centimetre in size. The rock now breaks along very irregular surfaces and is known as a *gneiss*. Ultimate metamorphism will start to melt the rock to make a *migmatite* (described in Chapter 4).

Different compositions of starting materials will give the same general series of rocks. However some igneous rocks, sandstones and limestones have the wrong compositions to form the sheet silicate minerals that give the *slatey cleavage*. Instead they tend to form a more granular texture; the limestones forming marble, the sandstones forming quartzite, and the igneous rocks a variety of compact materials full of interlocking needle-like crystals (fig. 6.13) (overpage).

Some rock metamorphism is not associated with orogeny, but has formed due to heat without pressure; for example the heat supplied by a nearby large body of magma. Around the Cornish granites the sediments have been altered and the minerals changed, but without the squeezing of the *orogenic metamorphism* and so the rocks are filled with new minerals but lack the slatey cleavage or *schistosity* that only comes with pressure. This metamorphism due to heat alone is known as *contact metamorphism*. It is easy to differentiate, because it can be shown to increase in degree towards the intrusion and also to ring it (fig. 6.12).

6.13 Metamorphism – Sandstone (a) to Quartzite (b) – Limestone (chalk) (c) – to Marble (d) – Basalt (e) to Amphibolite (f)

(a)

(b)

(c)

(d)

(e)

(f)

them, are 8,000m of 'Dalradian' sediments. The lower portion of the Dalradian continues the marine sediments of the Moine, while the upper part of the pile contains volcanics with pillow structures and turbidite sediments. We now know that this pile was laid down right across the pre-Cambrian/Cambrian boundary as fossils have been found in some of the rocks. However the rocks have been folded, large areas are the wrong way up, as well as metamorphosed, so it is only through recent patient and meticulous

unravelling that the Dalradian has been understood.

The Moine and the Torridonian formed at the same time, between 1,000 and 800 million years ago. There then follows a gap before the Dalradian was deposited about 600 million years ago. During this 200 million years, there was slight folding and metamorphism on the western margins (fig. 6.15).

Thus the late pre-Cambrian history of either side of the Iapetus ocean, can be elucidated. However, on the northern margin of this ocean, unlike

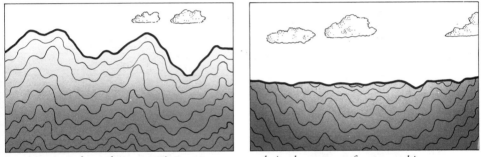

6.14 Erosion of complex mountain structures to reveal simpler pattern of metamorphism

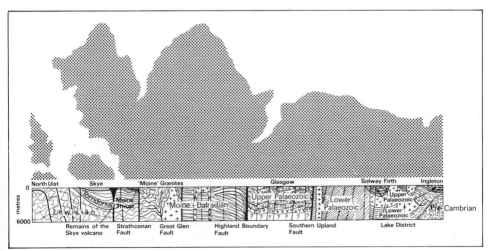

6.15 Cross section through Northern Britain

73

6.16 *An outcrop of Lewisian gneiss*

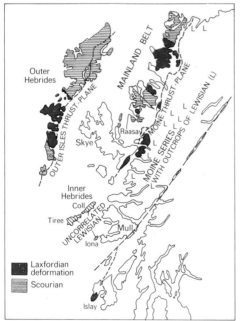

6.17 *The outcrop of the Lewisian*

the southern, we have a chance to see what lay beyond and beneath – a question about which we can only speculate when thinking of England and Wales. The area of typical continental shield to the north west of Scotland, the '*Lewisian*', is a remnant of a series of former orogenies and mountain building episodes. This old continental crust forms the irregular coastal fringe to the North West of Scotland, as well as nearly all the Outer Hebrides. All the rocks are very much altered, having suffered high grades of metamorphism, often on more than one occasion. Some of the history can be unravelled using conventional geological techniques found from mapping; dykes cutting older rocks or one direction of folding superimposed on an older one.

However it is only through *radiometric dating* that the story of these, rather uniform, banded, grey gneisses (fig. 6.16) has come alive; within the laboratory rather than out on the rocks themselves.

Two major episodes of orogeny are defined. The first gives dates from 2,900-2,200 million years and is known as the *Scourian*; the second from 2,200-1,500 million years is the *Laxfordian* (fig. 6.17). Between the two, there was a period of intrusion of dykes full of basaltic magma running roughly NW-SE, that are particularly prominent in the southern part of the Outer Hebrides and in the central portion of the mainland fringe. Much of the north of the Isle of Lewis and northern parts of the mainland were metamorphosed in the Scourian and then again in the Laxfordian. Of course, seven hundred million years is far longer than the duration of our more recent orogenies and it is likely that inside this time, one is actually recording several separate events.

For instance, some of the dykes suffered deformation, before other parallel ones were intruded. Despite the excellent outcrops of these rocks over the whole area, so much has been lost in erosion that we can never hope to work out anything other than a shadowy picture of events. One feature that is characteristic of all these old continental shields is that the rocks seem to have been reworked in every period of orogeny and that dyke swarms are very common. It is likely that the rigid plates and continents of

6.18 Fossil Lewisian landscape, 1000 million years old emerging from beneath the Torridonian sandstones that are now left as isolated mountains – NW Scotland

the earth did not then exist and that their antecedents were much more plastic. The rocks we see exposed today must have been 15 or 20km down, beneath the surface of this strange former world, at a time when the earth must have seemed like a different planet.

What were these rocks before they became metamorphic gneisses? No rocks older than 3,600 million years have been discovered on earth, yet we know from other radiometric information that our planet was formed, along with the other planets and meteorites of the solar system, some 4,600 million years ago. The rocks of the Scourian probably started as volcanic ashes or lavas, formed a hundred million years or so before they were metamorphosed. However, tracing back any kind of history, before the metamorphism obliterated everything, is bound to be highly speculative. The story has been lost with the erosion of the huge thicknesses of rocks that used to cover the area. The sediments produced from that erosion must be around somewhere. Possibly some pebbles from within them might help add detail to the missing sections of our tale, of North West Scotland in a youthful world.

At the time of the formation of the Torridonian sediments, the continent was sinking so fast that the new sediments covered over a complete landscape. Beneath the Torridonian it is now starting to re-appear again; mountains rising 900m above the valleys, of a fossil landscape (fig. 6.18), from a time about a billion years ago when there was no life anywhere on land and only the simplest of organisms in the sea. The terrain then was barren and rugged; not so very different from the bare rocks of some of the islands and coast in that part of Britain today.

The Pre-Cambrian is also officially known as Precambrian.

LIMESTONES

7

Mortal remains

the North and South Downs (fig. 7.1, 7.2 and 7.3). The scenery is a combination of rock and vegetation – the close cropped grasslands that have made the limestone countryside into an image of England. The 'white cliffs of Dover' made out of chalk limestone symbolise 'home' and may also have inspired the name 'fair Albion'. And all this for a rock that was once sufficiently insubstantial to be dissolved in water!

Very little water on earth is pure, for on contact with the materials of the rocks, certain elements tend to dissolve within water as *ions* (ions are atoms or molecules with an electrical charge, that can pair with water). Ions cannot be seen once they are dissolved, but they can be made to 'reappear'. In regions of Britain where the water supply has passed through the rocks before being collected, it is generally contaminated

While the rocks of Scotland, Wales, the Lake District and South West England have formed through a violent geology of deformation and orogeny, much of the softer scenery of England has come from the gentlest of rock formation processes – that of the limestone. Limestones make up much of the Pennines and the Derbyshire Dales, the Cotswolds, the Mendips and the rounder hills of the Chilterns and

7.1 Limestone country – limestone scarp, North Wales

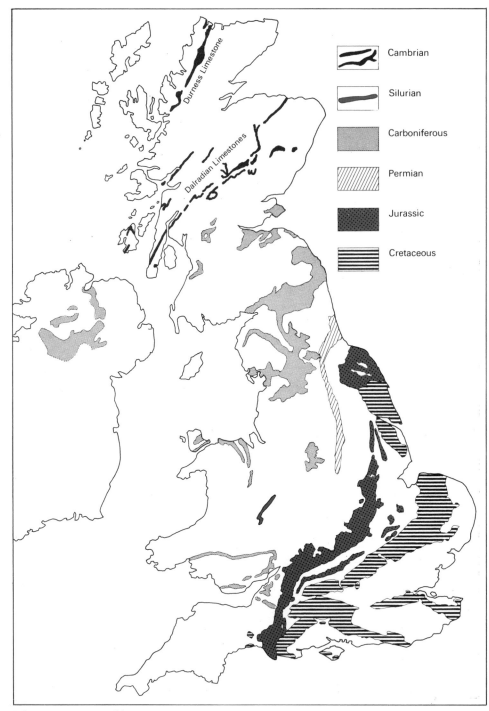

Legend:
- Cambrian
- Silurian
- Carboniferous
- Permian
- Jurassic
- Cretaceous

Durness Limestone

Dalradian Limestones

7.2 *Major outcrops of limestone in Britain*

7.3 *Limestone country – the chalk cliffs of the Seven Sisters, Sussex*

7.4 *Limestone formation!*

with quite harmless amounts of ionic impurities. It is dissolved calcium and bicarbonate ions that make hard water 'hard'. They can be made to re-appear, or *precipitate*, when the water is boiled away, as in a kettle (fig. 7.4).

A similar kind of process, of ions re-appearing out of solution, can, over long periods of time and under the appropriate conditions, be important enough to produce great thicknesses of rocks. One of the appropriate conditions is that the water is not full of mud or sand, that will settle out to form one of the sedimentary rocks of Chapter 2. Limestones form away from eroding land, in quiet water conditions.

Although there are several ways of making the ions of limestone re-appear out of solution, the biological world of the water has an interest in abstracting them in order to build hard and rigid pieces of skeleton or shell, that can protect and strengthen the organism. In general, therefore, the concentrations of the ions never build up sufficiently for limestones to form directly from the water. Instead, the animal and even the plant world borrows from the mineral world. On death, the remains of skeletal material sink to the bottom of the water to build up future rocks. The mineral most commonly used by organisms for these purposes is *calcium carbonate*. 'Carbonate' is an ion made up of three oxygens and one carbon atom, that forms part of a complex chain of chemical reactions, that also includes the carbon dioxide in the air. Through this series of reactions, there is a form of regulation of the amount of carbon dioxide in the atmosphere, that we hope will be maintained, in spite of the enormous amounts of carbon dioxide added by burning fossil fuels: coal and oil.

Marine organisms use three kinds of calcium carbonate to build their shells: *Aragonite* – a dense compact mineral form; magnesium rich calcite – that has a large number of magnesium atoms present where one would expect calcium; and ordinary calcite. Calcite is the only one of these that is actually stable, but the other two are easier for certain organisms to manufacture. The form in which the calcium carbonate is used by the organism can have a significant affect on the resultant sediment; as we shall see.

The amount of calcium carbonate that can dissolve in water increases as one raises the pressure and lowers the temperature. Under the oceans there is

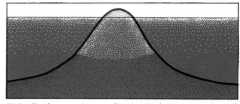

7.5 *Carbonate 'snow line'. Sinking particles of carbonate dissolve in the depths of the ocean and settle out only on shallower slopes of a Pacific island*

7.6 *Modern reef*

7.7 *Stromatolite – a fossil reef built by algae*

a kind of *carbonate 'snow-line'*, formed because the shells of the dead organisms will dissolve back into the deeper colder water towards the ocean bottom. As one heads away from the equator, the 'snow-line' becomes shallower, where the water gets colder. For the mid-Pacific this is about 3,700m (fig. 7.5); for parts of the Atlantic, due to a different water circulation, it is nearer 5,000m. Below these depths, carbonate sediments cannot form.

Most of the British limestones were formed in relatively shallow water during periods of stability. Most were made predominately from the remains of organic skeletal material, but in many rocks, because the calcium carbonate can pass so easily into solution and out again, only traces of these remain as fossils. However some contain 100% fossils; the largest of all being fossil *reefs*, that were built up as banks from the skeletal remains of millions of dead organisms. The reefs we find today in the tropics are built out of coral – fig. 7.6. In the past, other types of organisms such as sponges or even algae used the same principals of reef construction. (The algae do not actually have skeletons but bond the calcite in thick mats that then provide a base for the next layer of

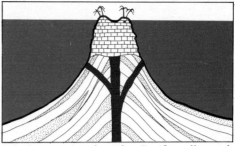

7.8 Cross section through a Pacific atoll – reef builds up on sinking extinct volcano

algal growth (fig. 7.7). Present day reefs, on top of extinct underwater volcanoes in the Pacific Ocean, may be more than 1,000m thick, for away from the mid-oceanic spreading ridge, the ocean floor beneath them cools and sinks. In order that the reef organisms can keep close to the sunlit water surface, they must keep growing on the backs of their dead ancestors, slowly building up an enormous tower of reef (fig. 7.8).

Reef limestones are found in the Silurian rocks at Wenlock, in Shropshire, where they grew in shallow shelf seas that fringed a deep

7.9 Shelly limestone

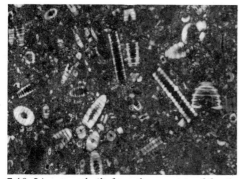

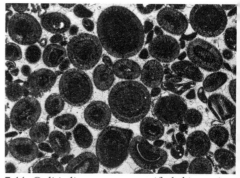

7.10 Limestone built from the remains of fossil sea lilies (crinoids)

7.11 Oolitic limestone – magnified thin section

trough that was filling up with greywackes in what is now Mid-Wales (fig. 7.3). In the Devonian of South Devon and in the Jurassic of Central England, lens shaped reefs formed as raised banks in a sea, that was slowly filling up with mud. The reefs had not only to keep close to the sea surface but also to stay above the rest of the sea bottom where the animals of the reef would become choked with sediment.

The golden age of British limestones was in the early Carboniferous Period, when the Devonian deserts sank beneath the waves, and limestones formed in the shallow seas. Some are made entirely out of shells (fig. 7.9); others are almost 100% made up of the stalks of *crinoids*, or sea lilies, animals with long articulated stalks, of calcite discs, that rooted them to the sea bed (fig. 7.10).

Many of the Jurassic limestones that make up the Cotswold Hills as well as some of those of Carboniferous and Silurian Age are made out of accumulations of small spheres, about 1-2mm in diameter known as *ooids*, from their ressemblance to caviar or fish roe. The limestone is termed *oolitic* (fig. 7.11). Each ooid consists of a series of concentric spherical shells,

each of which is bonded to the next by algal filaments. Ooids can be seen forming today in the shallow coastal waters of Florida and the Bahamas, always in fairly rough tidal conditions, for the spherical shape develops from the buffeting and disturbance. They may accumulate to form shoals just as if they were the sand that would be present if there were any erosion of material off the land. Instead, the sea rolls its own, alternative beaches.

After the end of the Jurassic period, much of Northern Europe became submerged beneath a sea in which up to 600m of the 'rock' we know as *chalk* was deposited. Chalk, unlike all the other limestones, is soft and crumbly. Under a very high power microscope (an electron microscope) it is possible to see the individual particles that make up the chalk (fig. 7.12). These are tiny ribbed shield-like plates, each one only 0.003mm across, known as *coccoliths*, that once formed the armour of the, only slightly bigger, spherical algal organisms known as coccolithoporids (fig. 7.13). They are still found today, about one hundred million years later, in large concentrations within the surface waters of the oceans. Sometimes more an a million of them can be found in litre of water. For this period during ne Cretaceous, the sea must have been oo deep, perhaps 200m or more, for nost reef or shoreline organisms and so nese algal remains have made up the najority of the calcium carbonate nud' that collected on the sea bottom.)ther fossils, shells and sea urchins are ommon.

Within the chalk there are horizons, or strata, rich in bulbous masses of very finely crystalline quartz, nown as *flints* (fig. 7.14). The quartz robably came from the skeletal emains of sponges, that choose to nake their framework out of silica nstead of calcium carbonate. The exact vay in which flints formed is still nknown, but in all the soft rocks of he south eastern part of England, they stand out' as the only hard material. 'rehistoric man even mined flints in order to manufacture stone implements fig. 7.15). Because they are so very ine grained, they fracture along curved onchoidal fractures like a glass. Where wo of these curved surfaces meet, it is

7.14 *Dark nodules of Silica (flint) within the chalk*

possible to make a very sharp edge. Most rocks will break unevenly around the mineral grains and can only be made as sharp as the smallest grain. At the Langdales, in Cumbria, prehistoric men made stone axes out of a fine-grained altered volcanic ruff that breaks like the volcanic glass *obsidian*.

The difference between the crumbly chalk and the well cemented limestone came about because of a difference in

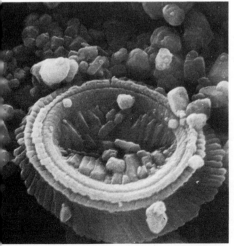

7.12 *Coccolith in chalk – electron micrograph*

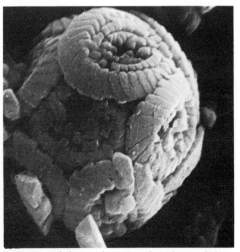

7.13 *Coccolithoporid armoured with coccoliths*

7.15 A primitive flint implement

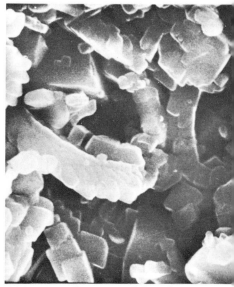

7.16 Recrystallizing coccolith – electron micrograph. Chalk turning into limestone from North Sea drill core

the form of the calcium carbonate. Coccoliths are made from calcite, while most other shell and skeletal fragments are made of aragonite or magnesium rich calcite. Unlike most limestones the calcite particles have not had to breakdown and recrystallise. Instead, all the particles have stayed separate – and the rock remains crumbly. In fact, North Sea Oil exploration has shown that where these chalks extend into the North Sea, under a considerable thickness of later Tertiary sediments, the slightly increased temperatures at depth have started to recrystallise the coccoliths and fill in the gaps, making a much stronger rock, more like an ordinary limestone (fig. 7.16).

Changes taking place after the rock has been formed are also important in the production of *dolomitic* limestones, in which half the calcium ions of the calcite have been replaced with magnesium to make the mineral dolomite. Strong brines, associated with a hot sun and a nearby sea, percolate through newly formed reef or beach limestones, giving solution and recrystallisation that hardens the rock and also alters the composition. Dolomite is more resistant to further attack than limestone and makes a better building stone.

Around some coral atolls and hot desert coasts, limestones form in the intertidal region, partly through the evaporation of sea water, to make a very hard rock that can grow fast enough to have included bottles and even skeletons from the Second World War – some of our most recent fossils.

Many of the limestone outcrops have been quarried, often for *cement*. Cement is made by heating up calcium carbonate with silica to make highly unstable calcium silicates. These break down with water to form new kinds of hydrated minerals, that bond

84

Continued on page 87

Where a sea becomes enclosed and the water evaporates faster than it is replenished, limestone and dolomite may appear early on in the succession of mineral precipitates, which form as the brine gets more and more concentrated and the water can no longer cope with all its dissolved ions. A succession of minerals form as the water evaporates; calcium sulphate after the carbonates, followed by sodium chloride and finally potassium chloride. Such conditions exist today around the Dead Sea in Israel and in some of the salt lakes of the Western USA (fig. 7.17). In Permian times, the then 'North Sea' became cut-off and dried up on several occasions to give rise to what are known today as the Zechstein Sea deposits, with up to 600m of '*evaporites*', mostly rock salt, but also potassium chloride – (mined at Whitby). To the western margin of this enclosed sea, only the first part of the evaporation is seen. As the basin dried up, the water retreated to the deeper parts of the basin to the east. The presence of overlying brines caused some of the Magnesian Limestones of Durham to suffer considerable changes – sometimes to develop textures like piled cannonballs (fig. 7.18). This limestone,

7.18 Cannonball Limestone from Durham

7.17 Controlled evaporation of sea water to make salt – Canary Islands

becomes progressively more and more unfossiliferous as one reaches the top of the sequence where the water, that

7.19 *Thick evaporite salt deposits mined in Cheshire*

formed them, became too salt rich for life. A similar evaporating basin existed in parts of the Midlands during the Triassic and provided 250m of salt deposits that are now mined in Cheshire (fig. 7.19). These evaporating basins must have filled and dried up on several occasions, for a thousand metre column of sea water can only produce 15m of salt!

Another important rock, formed from the separation of ions out of water is *ironstone*. Iron becomes enriched through being weathered out of rocks, dissolved in solution and later deposited, generally in lakes or rather stagnant lagoons, as a mixture of iron oxides, hydroxide and carbonates. Many of the Jurassic ironstones (still mined) (fig. 7.20) that outcrop from the Cleveland Hills in Yorkshire, right down to Oxfordshire, are oolitic, some have even replaced earlier ooids that were made of calcium carbonate.

7.20 *Jurassic Ironstone quarry in Northamptonshire*

7.21 Limestone cavern with roof stalactites
and floor stalagmites
7.22 Limestone: collapse of caverns to form
a gorge as at Cheddar (inset)

together sand or gravel to make the
man-made rock, concrete. Limestone is
also a convenient building stone; the
Pyramids, the Houses of Parliament
and many of the buildings of Oxford,
are made from limestones. Their one
enemy is acid; particularly modern
sulphur dioxide rich industrial

7.23 A typical dry valley in the chalk

where the streams have dissolved caverns along which they can travel below ground (fig. 7.21). When the roofs of several interconnecting limestone caverns collapse, a gorge may form, as at Cheddar (fig. 7.22). Where solutions drip and evaporate on a cavern ceiling, the calcite may reappear in *stalactites* or on the floor as *stalagmites*. A poorly formed limestone rock, called tufa, can be seen around waterfalls or in the top layers of some river soils, formed partly through evaporation and partly through organic activity. In the weird cracks and joints of the *limestone pavement* (fig. 7.24), we can clearly see that the stone that comes from the water all too easily goes back to rejoin it again.

pollution, which turns rain water into a weak acid that can easily attack and corrode both calcite and dolomite. Even the hills made out of limestone are subject to a similar chemical erosion. The dry valleys of the chalk and limestone country (fig. 7.23) mark

7.24 Chemical erosion – limestone pavement above Malham Cove, Yorkshire

FOSSIL FUELS

8

Potential energy

The more we learn about the other planets within our solar system, the more extraordinary our own planet becomes. Most remarkable is that at the places where the rocks, the atmosphere and the water meet, there is nearly always life. There are the plants that function by building complex organic molecules from nutrients of the earth in combination with sunlight, water, and carbon dioxide, and the animals that live on energy contained in the organic matter of the plants. This living world, resting on the surface of the earth, may seem to have little to do with the rocks beneath. However, under certain, rather special circumstances the dead remains of animals and plants may become a part of the rock world in their own right. If some of the organic carbon–hydrogen–oxygen molecules of the plants and animals are partially or totally preserved, in the condition in which they existed while alive, then they will also take with them some of their 'stolen' solar energy. The organic

8.1 Cutting peat, Galway

8.2 Mangrove swamp – coal of the future

molecules were originally built up by breaking down carbon dioxide and water – reactions that require energy to power them. To regain this energy we have to recombine them once again with oxygen. Thus the most important condition for the formation of a *fossil fuel* is that it should build up away from the presence of oxygen. This is possible in stagnant water as is found in swamps or bogs.

Peat bogs, formed through the accumulation of dead plant remains underwater, can be seen in many parts of the British Isles, notably Ireland (fig. 8.1). Most of them have only been built up since the last ice advance, within the last few thousands or tens of thousands of years. To see material being formed today that may become part of the future rock world we can look at the Fens of East Anglia or the Somerset Levels, where, at the bottom

of the pile, the peat is gradually losing water and starting on the first stage of the journey from peat to coal.

Peat deposits today are forming in two world-wide belts, the temperate one (as in this country) and a tropical one, that is separated from the temperate one by the 'arid zone'. It is the tropical one of mangrove swamps and steamy jungle that appears in our imaginations when we think of the origin of the British coal (fig. 8.2). Along the East Coast of the USA are coastal swamps, the Everglades and Dismal Swamps, that may be potential coals of the future. These are protected from the powers of the storms and the sea by sand bars (fig. 2.10).

Although coals were formed as far back as the pre-Cambrian (made out of the concentrated remains of algae, but only in very thin seams) the coal-age in Britain was sufficiently remarkable to

8.3 *German brown coal quarry at Bergheim an der Erft*

cause the period in which it occurred to be called the 'Carboniferous'.

The transformation of peat into coal is really one of metamorphism, it is just that the organic material is much more sensitive than rocks. Once the peat is buried about 10m down, the microbes which decompose the organic molecules die and further changes in the chemistry and substance take place more slowly (fig. 8.4). The temperature at which these early breakdown reactions occur is most important. German *brown coals*, formed 40 million years ago, are more decomposed and consequently more finely grained than those formed only 20 million years ago when the climate was colder. These thick German brown coals formed on the edge of deltas associated with the Rhine rift valley, that continued sinking and therefore preserved them (fig. 8.3).

Peat alters to the softest of brown coals when it becomes buried beneath about 200-400m of overlying sediments. The peat of some of these delta deposits must have originally been up to 100m thick, but through a considerable loss of water they have been compressed to about one third. To alter to a *bituminous coal*, like most of the British coals, requires compression by about one half again. A metre of hard coal may represent about 6m of original peat. This transformation requires between 1,500-4,000m of overlying rock. The changes from a peat to the hardest of coals are measured in terms of *rank*. High rank coals, or coals that have been deeply buried, are richer in carbon than low rank coals, because they have lost most of their hydrogen (and water etc.).

However it is not the pressure, but the temperature and the time that are

8.4 The effect of increased pressure on naturally occurring carbon rich rocks (a) Peat, (b) Brown coal, (c) Coal, (d) Graphite, (e) Diamond

more important in *coalification*. Further heat may turn a *bituminous coal* into an *anthracitic coal* that is very hard and carbon rich, like those found in some parts of the South Wales coal-field. With still higher temperatures, the carbon will start to form crystals of *graphite*, once mined near Keswick, Cumbria for the manufacture of 'lead' pencils. At the very highest pressures, carbon forms *diamond*, a strange transformation from soft black material into the colourless hardest natural mineral.

Natural diamonds have been forced up by gases along pipe-like holes from even deeper down than the volcanic magmas, perhaps 200km underground in the Upper Mantle. Unfortunately, these pipes are only found in the middle of old continental regions and Britain is not one of these. The carbon of diamonds was unlikely ever to have been plant material; instead it is the carbon that is still leaking out from deep down in the earth, where it has been since the earth's formation.

The rate of peat formation is about 1mm/year for temperate bogs rising to up to 3-4mm/year in tropical swamps. In N.W. Borneo 17m of peat has formed during the last 4,000 years, from luxuriant mangrove vegetation. At these rates, a soft brown coal may represent 1,000-2,000 years per metre, a bituminous coal nearer 6-9,000 years.

Apart from a few thin coals of the Jurassic age, all the British coals are Upper Carboniferous (fig. 8.6). Throughout the period there was uplift of some of the Highlands of Scotland and of a landmass that stretched across the South Midlands. This was sufficient to feed sediments to the rivers and keep the deltas in existence, despite continued subsidence. The *coal seams* occur with sandstones, mudstones and sometimes even limestones in constantly repeating cycles that reflect a battle between the land and the sea. After a peat deposit had formed for a period on land, the sea would flood over the sinking ground depositing mussel beds and, if there was no sediment flowing in, limestones. Then, as the land gradually reclaimed the area, first mud was deposited that gradually 'coarsened up' into sand as the river mouth came closer. Finally sand shoals appeared, that supported the first plants, their roots passing through the shallow water into the ground below (fig. 8.5). The plants leached the sediments below of all soluble elements leaving a residual sand of almost pure silica. This bleached rock, found beneath coal seams is known as seat earth or *ganister* and because of its purity, has been used

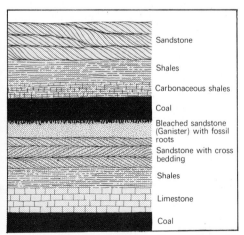

Sandstone

Shales

Carbonaceous shales

Coal

Bleached sandstone (Ganister) with fossil roots

Sandstone with cross bedding

Shales

Limestone

Coal

8.5 Typical cycles of deltaic sediments from the British coal measures

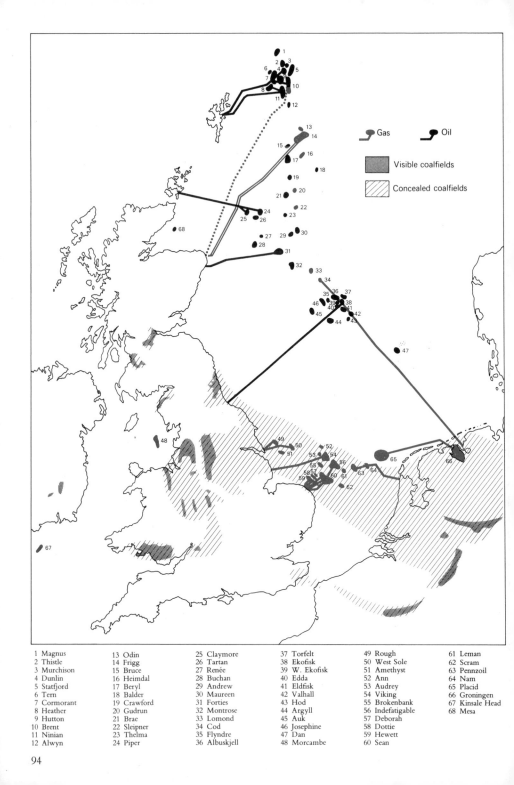

Gas   Oil

Visible coalfields

Concealed coalfields

1 Magnus	13 Odin	25 Claymore	37 Torfelt	49 Rough	61 Leman
2 Thistle	14 Frigg	26 Tartan	38 Ekofisk	50 West Sole	62 Scram
3 Murchison	15 Bruce	27 Renée	39 W. Ekofisk	51 Amethyst	63 Pennzoil
4 Dunlin	16 Heimdal	28 Buchan	40 Edda	52 Ann	64 Nam
5 Statfjord	17 Beryl	29 Andrew	41 Eldfisk	53 Audrey	65 Placid
6 Tern	18 Balder	30 Maureen	42 Valhall	54 Viking	66 Groningen
7 Cormorant	19 Crawford	31 Forties	43 Hod	55 Brokenbank	67 Kinsale Head
8 Heather	20 Gudrun	32 Montrose	44 Argyll	56 Indefatigable	68 Mesa
9 Hutton	21 Brae	33 Lomond	45 Auk	57 Deborah	
10 Brent	22 Sleipner	34 Cod	46 Josephine	58 Dottie	
11 Ninian	23 Thelma	35 Flyndre	47 Dan	59 Hewett	
12 Alwyn	24 Piper	36 Albuskjell	48 Morcambe	60 Sean	

94

for the manufacture of silica glasses. Even within the coal seams it is possible to trace cycles; this time cycles of vegetation. The first plants to grow were large trees that could survive with their roots passing through the shallow water. The high concentration of woody material gives rise to shiny bright coals. As the swamp emerged from the water, smaller bushes and plants replaced the big trees and the coals become duller and richer in the remains of spores. The spore remains from some layers show that there were periods when the forest was made up of only one kind of plant. As the peat was submerged, the big trees once again became more important.

Often seams have many cycles of coal type within them. Sometimes there are charcoal rich layers of burnt material, evidence of fossil bush fires. The top layers may change to a carbonaceous shale as muds re-enter and sometimes the stagnant waters overlying a peat layer deposited ironstones. If the water was brackish, then sulphates within the brine could be decomposed by bacteria in the peat to make the coal sulphur rich. The amount of ash left in coal after burning depends on how much sediment entered the peat swamp with the water. British coals are fairly ash rich, the German ones ash poor. The rivers at present running across the peat in Florida have such a slow flow that all the sediment is dropped before entering the swamp.

British coals are found to the north and south of the Carboniferous Wales–

8.6 *Fossil fuels around Britain and the North Sea*

8.7 *Changes in the coal seam; 'splits', 'fade-outs' and 'wash-outs'*

South Midlands–East Anglia land mass; all around the Pennines; and in parts of the Midland Valley of Scotland. The surface outcrops continue for a considerable distance underground; the concealed part of the Yorkshire–Nottinghamshire coalfield has a greater area than that exposed and contains the greatest known reserves of any British coalfields. The North of England seems to have been an area of considerable subsidence, towards which the deltas piled their sediments making as much as 1,700m thickness of coal measures in Lancashire. The individual coal seams are generally one or two metres in thickness, (though many smaller ones are not mined), with a maximum of about 10m. The seams themselves may split in two or just fade away (fig. 8.7), just as one might expect from the pattern of rivers and swamps distributed around a delta.

Some of the English coal seams hidden below much younger sedimentary rocks have turned up in rather unexpected places. The Kent coal-field is buried below 300m of Cretaceous and Jurassic sediments, and continues under the channel into North France where it is also mined. A huge coal seam has only recently been discovered beneath the Jurassic clays north west of Oxford. The coal under the Channel is unlikely ever to get mined, although some of the Durham and Cumbrian collieries do extend under the sea for a kilometre or two. Fortunately we can tap some of the energy of another series of enormous coal seams, that formed under the North Sea, in a different way.

The coal of the North of England continues out across the North Sea into North Holland and Germany. In the offshore region to the east of Lincolnshire and to the north of Norfolk as well as offshore and onshore in Holland, methane *gas* has poured off the coals and collected in natural underground *reservoirs* (fig. 8.5). Natural reservoirs for oil and gas form in *porous* rocks that are sealed off with *impermeable* material. This section of the North Sea kept subsiding after the Carboniferous, and in the ensuing desert environment of the Permian sandstones were laid down, known as the *Rotliegendes Sandstone*. Following this, the *Zechstein Sea* flooded the area depositing shales and evaporites, that gave the sandstones an impermeable seal or cap. Minor amounts of folding and faulting have created dome-shaped structures in the sandstones, that have acted as reservoirs (fig. 8.8). Sometime after the beginning of the Cretaceous

the coals that lay below lost enormous volumes of methane, as they increased in rank, to become anthracitic. As with the high rank coals found in South Wales, the reason for this is not immediately obvious. But whilst the South Wales ones may have suffered heating due to nearby intense folding and faulting at the end of the Carboniferous orogeny to the south, the North Sea coals were probably heated up by magma intrusions connected with the early attempts to open up the North Atlantic at the end of the Cretaceous peiod. The coal seams that were the source for this gas are known to be very thick and to cover an enormous area. Unfortunately, without some major technological advances in coal extraction, it does not look as though they can ever be mined.

Further north in the North Sea there are reservoirs similar to the Permian sandstones that have also been filled from below, but this time with *oil* (fig. 8.6). Oil has a completely separate origin from the coal and the gas – although some oil can be forced out of

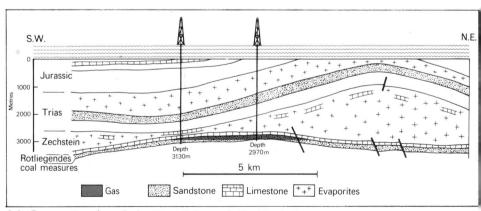

8.8 *Cross section through the West Sole gas fields, North Sea*

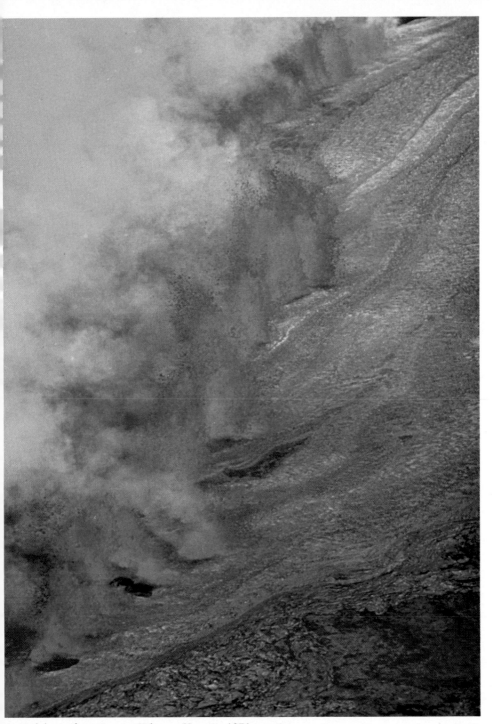

1 *A dyke in formation on Kilauea, Hawaii, 1971 eruption*

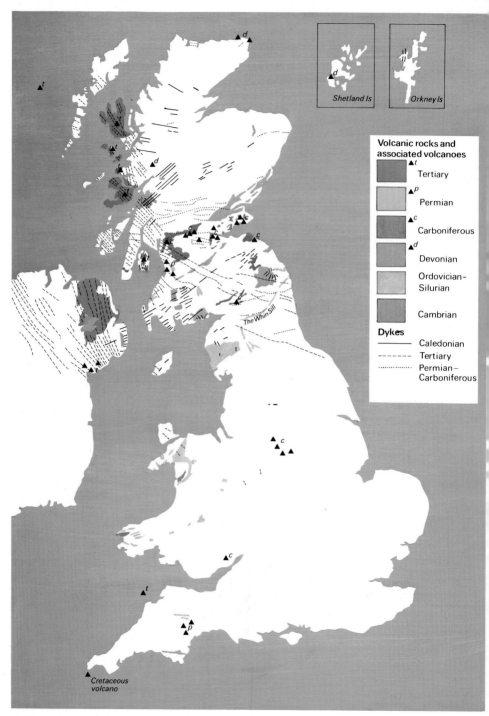

Volcanic rocks and
associated volcanoes

▲t
Tertiary

▲p
Permian

▲c
Carboniferous

▲d
Devonian

Ordovician-
Silurian

Cambrian

Dykes

Caledonian

Tertiary

Permian–
Carboniferous

The Whin Sill

Shetland Is

Orkney Is

*Cretaceous
volcano*

2 *Map of exposed volcanoes and volcanic rocks of Britain since the pre-Cambrian*

3 *Shap granite* – porphyrite *with large alkali feldspar crystals*

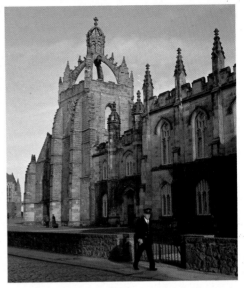

4 *King's College, Aberdeen* – granite in use as a tough building stone

5 *Crystal tableau – quartz with chalcopyrite*

6 Coloured natural quartzes polished for jewellery 7 Blue John vase – impure banded fluorite from Derbyshire 8 Cubic fluorite crystals in dolomite limestone

9 Magnetite – lodestone 10 Banded malachite (green) with azurite (blue) 11 'Native' copper
12 Galena (lead sulphide) 13 Chalcopyrite (copper iron sulphide) with dolomite
14 Fool's gold – pyrites (iron sulphide)

9

11

12

10

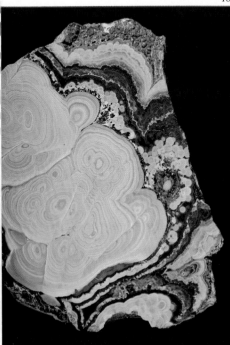

13

14

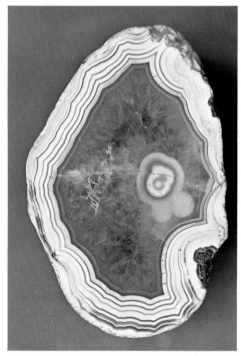

15 Banded agate nodule – (impure quartz)

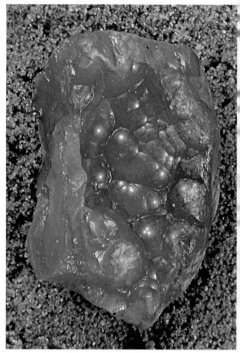

16 Chalcedony (a variety of fine grained quartz)

17 Polished pebbles from a Welsh beach

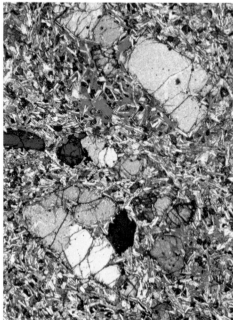

18 Enlarged thin section of an olivine basalt

In the west volcanoes are forming above a subduction zone on the edge of a contracting Atlantic Ocean.

To the east the sea advances and there is continued erosion and subsidence.

The south is affected by the continued plate movement of Africa into Europe which results in the squeezing up of the thick piles of sediments to form these mountains.

19 Britain 80,000,000 years A.D.!?

20 Torridonian sandstone mountains of NW Scotland resting on the ancient Lewisian basement including Spit Rock, Suilven, Cul Mor, Culbeag and Stack Polly in North West Sutherland

21 Duncansby Stacks near John O'Groats, Caithness – Old Red Sandstone

8.9 Kimmeridge, Dorset – oil pumping station

the bituminous coal seams. Oil comes from marine sediments – in a sense, oil is the marine equivalent of coal, but whereas on land one may have conditions where organic material is preserved in swamps, in the sea the dead animal and plant matter must be rapidly buried in mud at the sea bottom to have a chance of preservation, without becoming part of some other organism's food chain.

Most petroleum (the Greek 'spirit from the rock') is thought to have come from sediments that contained less than 1.5% organic material – not an unusually high amount for marine muds. The genesis and formation of oil involves a series of low temperature breakdown reactions in which, unlike the coal, the hydrogen stays behind, bonding with the carbon in hydrocarbons. The clay minerals themselves may have the property of storing the oil between the layers of their mineral structures, only to release it after a fairly considerable weight of overlying rocks has built up. Above all, the formation of petroleum must be slow and gentle. It cannot be hurried by rapid heating of the source rocks.

Shales and mudstones, rich in organic matter, of the Kimmeridge Clay (of Upper Jurassic Age), appear to be the *source rock* for all the North Sea Oil. Oil shales of this age are found on land in Lincolnshire and Dorset and oil is still obtained by pumping near the village of Kimmeridge in Dorset (fig. 8.9). The highest organic content of these shales appears to have been within the North Sea rifts that were very active at that period. While the source rock for the oil seems to have been, more or less, the same general sedimentary formation, the rocks that have provided a reservoir are incredibly varied. Some are even in formations that are older than the source rocks that might be expected, by the normal laws of stratigraphy, to

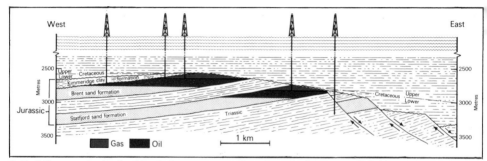

8.10 *Oil reservoirs in the North Sea – cross section through the Brent oil field*

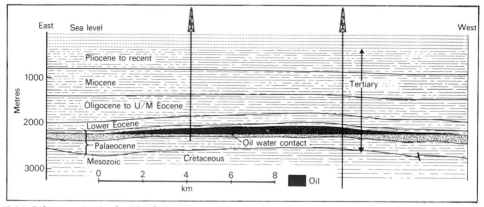

8.11 *Oil reservoirs in the North Sea – cross section through the Forties oil field*

be below them. However along the edges of the central North Sea rift valley, the Jurassic shales have sometimes been dropped by faulting below adjacent older sediments (fig. 8.10).

Suitable sandstone reservoir rocks have formed at various periods, when uplift adjacent to the North Sea has provided a coarser sandier sediment (fig. 8.11). The Forties and Montrose fields lie in sands of the Tertiary Age, trapped beneath later Tertiary shales, both in large anticlinal dome structures. Throughout the northern North Sea, there has been considerable recent *downwarping* that has allowed up to 3,000m of Tertiary sediments to

accumulate. Ekofisk and Dan oilfields are in fractured, porous, chalky limestone of Cretaceous Age, capped off with lower Tertiary shales. The Piper oil field is in Upper Jurassic coastal sand bars that have been forged into a reservoir through fault block movements; the Brent oil field is on a tilted fault block at the edge of the rift, in sands of early and middle Jurassic; and the Argyll and Auk fields also contain oil that has migrated into traps on the edge of the rift – this time of the Zechstein dolomites and Rotliegendes sandstones, that provided a home for the gas further south. In many parts of the North Sea, the *cap rock* of these reservoirs contains shallow folds that

8.12 North Sea oil rig

developed from faulting in the more brittle rocks below.

None of the North Sea oil fields is individually near the size of the massive Middle East ones. However, the variety of locations where oil has become trapped gives hope that more of these smaller reservoirs will continue to be found (fig. 8.12).

Our hostility to the oil pollution that has resulted from over-rapid exploitation sometimes makes us forget that oil is actually part of nature too. Natural oil seepage into the sea and even cliffs that catch fire are a feature of the Dorset coast. Natural blow-outs and oil fires have been found in the Middle East (fig. 8.13) and there are even huge reservoir formations that, a long time in the past, were eroded to release millions of tonnes of oil into the neighbouring rivers; that must eventually

8.13 A natural oil fire in the Middle East

have covered the sea over hundreds of square kilometres. Even so our perspectives in all these things, in pollution and in the conservation of these dwindling resources of fossil fuels must be brought into question because man has operated more than a million times faster than Nature would have done if left to herself. And most processes if speeded up a million times will at some point overload.

GLACIATION

9

After the ice

The rocks that built Britain are, in a way, only half the story. We cannot always find a simple relationship between the rocks and the scenery, for many of the shapes of the valleys and the hills have come from the work of a recent sculptor – ice. The active work of changing the landscape still continues in the erosion of slopes and in the transportation of material from high ground to low by rivers. However, the river's erosion compared to that of a *glacier* is like a spoon in relation to a bulldozer. A glacier can carry enormous quantities of rock

débris over long distances and can, through its sheer bulk, carve out huge troughs – some Norwegian Fjords have near vertical sides and are 2,500m from top to bottom (fig. 9.2).

Accompanying the glaciers there is the action of *freeze-thaw* when water gets into rocks, freezes and expands and then remelts penetrating even further into the rock along the cracks. This power can be seen in a more domestic setting (fig. 9.1), if a bottle has been filled full of water, sealed, and put in the icebox (not recommended) or if the water in a car radiator has frozen and cracked the pipes.

At the beginning of the last Ice Age the amount of snow falling on the British mountains in the winter became more than the amount melting each summer. It has been estimated that the climate need have been only a few degrees colder than it is today. Even now the estimated lower limit of permanent snow fields in Scotland is, at about 1,500m, only just higher than

9.1 *Ice power – the water to ice expansion that lifts a bottle top can fracture a rock*

9.2 *An ice carved valley – Geiranger fjord, West Norway*

the highest Scottish peaks, and on both Cairngorm and Ben Nevis there are two or three isolated gullies that can carry snow all the year round.

At that time, as the snow built up on the high ground, the lowest levels started to change and compact, rather like the way that a soft mud changes into a rock. The snow crystals re-crystallise and the air trapped in the snow escapes while the ice becomes solid and translucent. As the ice thickens, it will move down to lower ground. This movement is partly through slippage at the base and partly through changes in shape and flow within the substance of the ice, that behaves like a liquid of very high viscosity. A glacier is such a slow mover that it will fill up a large amount of valley; water may move on an average gradient 100,000 times as fast as ice. Because the ice is concentrated in such a thick mass it may travel, snake-like, to much lower ground, before melting completely; often to far below the snow line (fig. 9.3). The shapes carved by these valley glaciers can be seen in all the Highland areas of Britain (fig. 9.4), and the valley glaciers themselves can still be seen in Norway and the Alps. The glacial valley loses all the narrow ravine 'V' shape of an upland stream,

9.3 *A valley glacier and* 9.4 (inset) *a glaciated valley – Stob Dubh, Scotland*

as the glacier enlarges its channel to form a '*U*' *shaped valley*. (The 'U' shape allows it to contain the maximum volume of ice within the minimum surface and frictional area.) At the top and to the sides of the valley there is either an ice field that lies on and above the mountain peaks, or *cirques* containing small glaciers that lie below them (fig. 9.6).

Cirques form when small hollows on the fringes of the summit peaks become snow filled and suffer erosion from *frost-sapping* at the boundaries of the snow field, where *freeze-thaw* can work most effectively, and from *ice-plucking* once the snow has built up sufficiently to form ice. These hollows bite back into the mountain and when they are filled with glaciers of their own, these may *overdeepen* the floor to form a hollow. Since the ice melted these have filled with water to become small lakes or tarns, and are common in the Lake District and North Wales (fig. 9.5). The mountains themselves have been shaped by these *cirques* (fig. 9.10) (also known as *corries* or *cwms*) that may scoop out the rock like bites from an apple to leave a central pillar as at the Matterhorn (fig. 9.7). Where these cirque glaciers join a valley glacier or when one valley

9.5 A cirque with tarn – Helvellyn, Cumbria

9.6 *Cirque glaciers – The Alps*

glacier joins another, there is no need for them all to reach a common base level. To make the ice flow all that is needed is a downhill gradient for the *upper surfaces (fig. 9.9). As a result, hanging valleys* form, that were once all ice filled to the same level. Now emptied, the sudden breaks in gradient provide some beautiful waterfalls (fig. 9.8).

Lower down the valley, where the glacier melts, the weight of the overlying ice and its thickness decrease nearer the *snout*, or end of the glacier, and so the amount of underlying rock erosion also decreases (fig. 9.11). Although the top of the glacier is running downhill the bottom will tend to develop an uphill gradient that, once the glacier has gone, fills with water to become a lake. The lakes of the Lake District, the lochs of Scotland and the fjords of Norway, have all formed

9.7 *Cirques biting into a mountain – The Matterhorn*

9.8 (Top) *Hanging valley, Yosemite, California and* 9.9 (inset) *Hanging valley in formation*
9.10 (Above) *Cirques biting into a mountain – Slieve Donagh, Co. Down*

through this process of valley *overdeepening* (fig. 9.12), that may also be aided by the presence of harder rock bands cutting across the valley floor.

These valley and mountain glaciers, that can still be seen in Europe, are not the whole story of the Ice Age in Britain. For at times the ice was not just restricted to the sides of mountains and their neighbouring valleys, but formed a giant *ice-cap* or *ice-field* extending from Scandinavia, right across the North Sea over Scotland and, at its maximum extent, right down to Southern England (fig. 9.14). Such an ice-cap exists today only on Greenland

and Antarctica (fig. 9.13). In both continents the ice is more than 3,000m thick towards the centre. We can only surmise the sort of processes that are taking place beneath them, but we can find details from the land surface of N.W. Europe that was covered by such ice masses up to only a few thousand years ago. Over the Gulf of Bothnia and the North Baltic Sea there may have been between 3-4,000m of

9.11 Overdeepening valley in formation (right)

9.12 Overdeepened valley – Coniston Water, Cumbria (below)

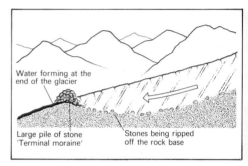

Water forming at the end of the glacier

Large pile of stone 'Terminal moraine'

Stones being ripped off the rock base

ice. The area is still rising as a result of the loss of this overlying weight. Some parts of Sweden have risen by up to 300m and still rise at a rate of a metre every century.

The 'Ice-Age', as it is known, actually consisted of several periods in which such ice caps advanced and retreated. It spanned the last two million years and still continues now. Within this period of time it has been possible to date the major ice advances from fossil, climatic and radiometric evidence within the kinds of sediments deposited at the fringes of the major ice fields. Only the last few hundred thousand years are known with any great accuracy, but it is within this period that the age of ice had its greatest impact on Britain.

A fairly sudden cooling can be detected from the kinds of fossils left in the sediments laid down around Norfolk 2 million years ago, followed by a warmer 'season' some 500,000 years later. But, as each ice advance destroys all the evidence of any lesser ice advances that preceded it, we know the story of the last 400,000 years, chiefly, because that marked the greatest ice advance of all. The ice sheet that spread out across the North Sea from Scandinavia was augmented by an ice cap based on Scotland with lesser ice-producing centres in the mountains to the West of England and Wales. It is likely that the whole of Scotland was covered by ice to such a depth that only the mountains at its very fringe, such as the Cuillins on Skye (fig. 9.15) protruded above it. There is a general lack of corries biting into the mountains to produce sharp pointed peaks in the Highlands, but many of them in the lesser ice fields in Skye and further south in the Lake District and North Wales, where ice may have failed to cover all the summits. In Norway and Sweden, the mountains are also rounded, whilst in the Alps, where the ice cap never

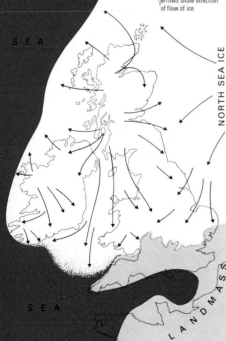

9.13 *The Ice Cap* (above left)
9.14 *Maximum extent of the Northern European ice cap* (below left)

ickened to blanket the highest peaks, the mountains are jagged and serrated.

Instead of following the valleys, the ice flowed outwards from the highest point of the ice cap. Given a downward gradient at its surface, the ice could even go up over hills that lay beneath. Blocks of rock from the Swedish mainland were carried up over mountains almost 2,000m high, before being deposited again on the Norwegian coastal plain. In flowing over a ridge of this kind, the ice will speed up and may tend to scoop out channels for itself to compensate for the restrictions to movement imposed by the hills. These old channels are bowl-shaped and now form lakes lying roughly parallel to the direction of ice flow. They are found on the Norway–Sweden border and also in Donegal, Ireland, where the ice sheet moved over a hard quartzite belt of rocks (fig. 9.16).

Where the ice flowing east off the Highlands flowed over Castle Rock in Edinburgh (fig. 9.17), the hard

9.15 The Cuillins, Skye – glaciated jagged peaks, remains of the Tertiary volcano (left)

9.16 Donegal Lakes, Ireland – glacier scooped hollows formed where the ice sheet travelled over a resistant band of quartzites (bottom left)

9.17 Castle Rock, Edinburgh – crag and tail formation. Resistance to the movement of the ice sheet from the hard volcanic neck of Castle Rock has caused the ice to scoop out compensating hollows (now Grassmarket and Prince's Street Gardens) (below)

volcanic plug impeded the ice, that had to find new compensatory routes and so carved out channels for itself to either side, now occupied by Princes Street Gardens to the North and Grassmarket to the South. The hard plug also protected the rocks immediately beyond it, so forming the ramp of High Street. This feature is known as *crag and tail* and is very common, often on a much smaller scale.

9.18 Scratches carved by rocks suspended in the ice

As the glacier flowed, rocks carried by the ice scratched the underlying rock surface leaving parallel lines, like *slickensides*, that give the direction of ice movement (fig. 9.18). The rate of flow of these massive ice cap glaciers is much faster than that of the little mountain glaciers and their erosional power enormous. However, away from the supply grounds, the ice stops eroding and starts depositing all the material captured from nearer its source.

The greatest deposition occurs at the *snout* or edge of the ice sheet, where much of the already melted ice washes out. A bank or mound forms, marking the limit of the ice, known as a *moraine* (fig. 9.19).

9.19 A terminal moraine – Glen Roon, Scotland

The moraine is made up of material dropped by the melting ice. During a period of ice retreat, the ice does not actually move backwards, but the point at which it melts gets nearer the source. The moraine follows, increasing in size wherever the ice boundary stays in place for a number of years. Any subsequent advance will destroy these *terminal moraines*. Other banks called *lateral moraines* form along thes edges of valley glaciers, made from material not ground off the rocks below, but undermined from the valley walls

higher up, nearer the source of the glacier.

The furthest advance of the British glaciers left a terminal moraine that can be traced across Essex, through Hampstead to Bristol. To the north of this line there are sometimes thick deposits of up to 60m of *boulder clay* from former or subsequent moraines. Even further to the north, where the ice must have been considerably thicker, this unconsolidated material beneath the glacier has been fashioned into elongated streamlined hills known as *drumlins*. The ideal drumlin is about a kilometre long and 20-30m high. Drumlins occur in flocks in Northern Ireland, between the Lake District and

9.20 Drumlins

9.21 Boulder clay – glacial till from former moraines

the Pennines and in several areas around the Southern Uplands (fig. 9.20). The exact mechanics of their formation is not well understood, but they ressemble, and may have a parallel with, sand dunes or sand moundlets amongst the ripples on a beach.

Glacial boulder clay, also known as *drift* or *till*, consists of a mixture of all sizes of fragments of rocks, for the glacier can carry everything and drops it only when it melts (fig. 9.21). The composition of this material depends on the erosional journey that the glacier has taken. The most interesting of the boulders are the *erratics*, rocks that have sometimes journeyed an enormous distance since they were picked up by the ice. As some types of igneous rocks are very local and very characteristic, the distribution of such rocks, as erratics, helps map the ice journeys. Shap granite pebbles travelled, at different times, right the way across the Pennines into Norfolk, whilst on other occasions, rocks of rhomb porphyry, a volcanic rock with massive white feldspar phenocrysts, arrived all the way from Oslo. These erratics can not only provide a

9.22 Perched erratic boulder

fascinating story of the ice but have also allowed rocks of great interest to reach parts of the country otherwise poor in exotic pebbles. It has even been suggested the blue-stones at Stonehenge were local erratics, carried by the glaciers from South Wales! Sometimes, where a boulder clay was deposited on high ground, all but the largest, most massive boulders have been washed away, leaving strange perched rocks completely unlike the material below – the visiting cards of the glaciers (fig. 9.22).

Even more curious things than drumlin formation may be going on beneath an ice sheet. As the ice melts, the water carves itself channels and ice caverns, where it flows towards the edge of the ice sheet. These rivers under the ice carry the debris once held within it. Some of this debris becomes redeposited, much as it would in a river, only this river has banks of ice. Once the ice has melted the course of the river is left as a raised bank that may run for many kilometres. These are known as *eskers* (fig. 9.23). Eskers are fairly rare in Britain but in some parts of Sweden and Finland, where they are very common, the roads and railways run along them. Sometimes the rivers beneath the ice carve out channels for themselves in the soft material below. Two famous ones, lost beneath the drift in Suffolk and South Cambridgeshire, have very steep sides and like some eskers, wander up hill. Rivers beneath glaciers can run uphill! Under a considerable pressure of water, the river is completely enclosed as if in a pipe and will flow as long as the end is at a lower level than the beginning.

The melt water from glaciers may become dammed, for short periods, between the ice front and the surrounding hills, to form a temporary lake. At several occasions during re-advances of the ice in England, lakes formed; the largest being 'Lake

9.24 Parallel roads of Glen Roy – former beach deposits of ice dammed lakes

Harrison' that extended from Leicester to south of Stratford, and was dammed to the north by the ice. Other lakes formed at different times; one in the Vale of Pickering in Yorkshire and another along the Welsh borders. These lakes overflowed across ridges, eroding channels that can still be seen perched high above the plains that were the original lake floor. As the ice front moved, different overflow channels came into operation. The most curious of these *ice dammed lakes* formed in Glen Roy in Scotland and left beach deposits 15m wide on three occasions, marking different lake levels, from separate overflow channel heights, (fig. 9.24). These beaches run as horizontal 'roads' along the valley sides. Ice even altered the course of the Thames – it used to flow out at the Wash. The ice dammed it to the north and it was forced to cut across the chalk hills and join with the River Kennet, the original 'London's river'.

Away from the ice itself, Britain was suffering arctic tundra and permafrost conditions. In summer, streams flowed above the frozen ground even where the rocks below were normally

9.23 A British esker made from former sub-glacial river gravels

9.25 *The site of the Ice Age 'Niagara of Yorkshire' – Malham Cove*

permeable. The ice acted as a seal, making rivers run along the '*dry valleys*' of the chalk and limestone country. The cliffs at the end of Malham cove were a waterfall even higher than Niagara (fig. 9.25). Other features of tundra have been found; in parts of East Anglia that missed out on cultivation it is still possible to see *ice-polygons*, an arrangement of stones caused by the repeated freezing and thawing of the ground.

The other more widescale feature of the Ice Age was the enormous drop in sea-level that resulted from so much water being incorporated into the ice caps of the Northern Hemisphere. The ice-cap lying across North America was even bigger than the one lying

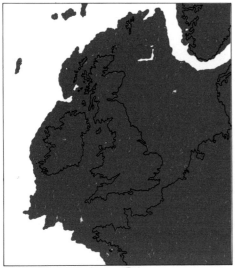

9.26 Britain's outline with lowest Ice Age sea level

9.27 If all the ice caps melted – 'the British Isles'

across Scandinavia. During the periods of low sea level, up to 135m below the present level (fig. 9.26), all the rivers carved for themselves valleys down to reach it. These had flooded to become estuaries when the sea rose to its present level, only 8,000 years ago.

The origin of Ice Ages is still controversial. Some cut-off in the amount of energy received from the sun seems a likely cause. Other Ice Ages are recorded in the rocks of other periods: one in the Southern Hemisphere between 320 and 280 million years ago and another one that formed thick deposits of boulder clay, since turned into a rock called *tillite*, across the Dalradian of the Southern Highlands at the pre-Cambrian/Cambrian boundary. The recent Ice Age may have only just begun; we are certainly still within it. The geological record shows us that the world at present is unusually cold; during most periods there were probably no ice caps present anywhere

on the globe. The fear of a new ice advance has often been suggested but just as frightening a political and economic spectre is the melting of the ice-caps. That would raise the sea level by 70 or 80m. During recent interglacial warm periods, higher sea levels (and rising land), formed several levels of beaches around Scotland; the highest about 30m above present sea level (fig. 9.28). Britain took on a new entity when the action of rising sea level and increased post glacial erosion separated it from France. Imagine the shape if this continued to its limit (fig. 9.27).

9.28 Raised beaches on Islay

BRITAIN'S PAST

10

A matter of latitude

So far in this book we have looked at the various ways in which the rocks were formed, the processes that they have suffered and finally how some of our landscape was shaped. In joining all these into a single journey of imagination, we can attempt to reconstruct a Britain through time. Of course Britain, as we know it, did not have any existence in the past. Only a short while ago there was land all the way to France. Further back in time,

the whole country may have been below the sea. As we showed at the end of the last chapter, a rise in sea level would completely alter the country's outline. The land/sea boundary, that is so strong in our visualisation, bears little relation to the geology, which continues under the sea. So instead of using a coastal outline we can pinpoint some major cities and use them as our reference points.

Most rocks of the geological record are rather unreliable as archives, because they tell us only about the conditions where they were laid down and little of the high ground from which they may have come. The scenery of the high ground is always more complicated than that of the low and always much faster changing. We can obtain only a general picture of the mountains from the kinds of particles

10.1 Ben Nevis in the Cambrian

and the size of fragments worn off them. The problem becomes more difficult when rocks of a certain age are missing from an area and we do not know whether they were laid down and later eroded or never laid down at all. Was there once chalk covering the whole of Wales and Ireland? We will never know as the evidence is missing.

We shall start the journey at the pre-Cambrian/Cambrian boundary; where our picture can have some hope of accuracy. At that time there was a divided Britain. 'Scotland' and 'England' were separated by an ocean. How far apart they were, we cannot say because the ocean crust that separated them is lost. However they were far enough divided to have distinct types of marine life, evolving in different ways, and swimming in their own coastal seas. The glaciation of the Ice Age immediately before (see Chapter 9) was remarkable as Scotland was in the tropics, slightly to the south of the equator. From the evidence of rocks from other continents, this Ice Age seems to have been world wide. As the sea level rose, after the ice had melted, limestones formed in the shallow shelf seas while further towards the ocean, the sediments of the Dalradian slowed down in their accumulation (fig. 10.1). Two or three thousand kilometres to the south, most of the Midlands and South and East England were covered by shallow seas; while to the north and south of a long finger-like, mountainous island that ran from South East Ireland through Anglesey up to Lancashire, there were deep sea trenches, filling up with sediments (fig. 10.2).

As the ocean began to close,

subduction zones formed at either side. The northern margin began to look something like the Andes while the southern one was a series of volcanic island arcs, more like Japan. On the northern margin of the ocean, through the Cambrian, Ordovician and Silurian periods mountains rose to the north and deep troughs formed immediately to their south. These, in their turn, were squeezed up into new mountain chains. England to the south and east was much more tranquil, with low islands and shallow seas. As the two sides of the ocean finally came together, long thin mountainous islands, running NE-SW, rose up out of the sea. At the final collision the whole area was lifted up to form a massive highland tract to the north, with smaller parallel mountain ranges across South Scotland, North England and Wales. The Caledonian orogeny was almost over.

10.2 Anglesey in the Cambrian

Britain still lay about 25° South of the equator. The ocean crust had almost entirely disappeared, but some of its melting products still bubbled up to help make the late stage granite intrusions. Volcanoes continued pouring out andesites through parts of Scotland; particularly in the Midland alley that was a desert basin next to he mountains (fig. 10.3). This was the ime of the Devonian 'Old Red andstone Continent'. For the first ime in our story most of Britain was ry land (fig. 10.4). Only the south of ngland was still under the sea, the oastline changing rapidly as deltas rom the north advanced and retreated n the sinking edge of other ocean asins in what is now central Europe. lash floods in the desert brought onglomerates and sands down off the nountains into the plains. The desert owlands became infilled. One of them xtended from the West Midlands long the Welsh borderlands into embrokeshire, one next to Anglesey, ne between the Lake District and the ennines, but the largest of all were urther north in the Scottish Midland alley and around the Moray Firth

10.3 Glasgow – looking north in the Devonian

and the Orkneys. Up to 10km of sediments piled into these *downwarping* basins, with pebble conglomerates forming near to the mountains and muddy lake flats further to the east.

The sinking of this land beneath the waves marked the beginning of the Carboniferous. At first the seas were clear and only limestone formed. The old mountains of the Caledonian had

0.4 Sheffield in the Devonian

Changes in past British climates can best be understood if we know where Britain stood on the globe at a particular time. The world climatic belts; the equatorial tropical belt and the symmetrical pairs of desert and temperate belts to either side are a product of the major atmospheric circulation pattern set up by the earth's rotation and presumably existed in the past. Through the movement of the earth's plates, Britain has wandered across the globe (fig. 10.5). Because the earth's magnetic field comes from rotational disturbance in the liquid core of the earth, it is oriented close to the earth's axis. Some rocks, when forming, contain tiny crystals of magnetite, that like little compasses, align themselves with the earth's magnetism. Through very careful measurement on the solid rocks, it is possible to work out the direction of this magnetism when the

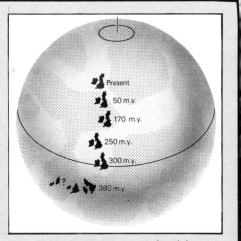

10.5 *Britain's journey across the globe*

rock was formed, and thus the rock's former position on the globe. The dip of the 'magnets' can give us Britain's past latitude. This study is known as *palaeomagnetism*.

been worn away, but later in the Carboniferous uplift began again for localised parts of the country and renewed erosion made the seas muddy. Limestone deposition ended, to be replaced by a long period of deltaic sedimentation.

During this period Britain sailed across the equator, starting at about 15°S and reaching about 10°N at the beginning of the Permian. The old story of some parts of the country rising, while others sunk, was repeated on a small scale throughout the Carboniferous. In the Northumberland trough the desert continued after the end of the Devonian. In the Scottish Midland Valley, alkali basaltic volcanoes developed with surrounding lagoonal flats (fig. 10.6). From Wales across to East Anglia there was land,

and to the south, shallow seas bordering a deep trough around Cornwall and Devon. This received little sediment until the second half of the period when the deltas covered much of the country – the rest being mostly mountains. The oscillations of the deltas with the sea have produced, in some localities, as many as 60 marine bands in the Coal Measures. Rivers flooded off the wide mountain ranges to the north west and off the smaller lower ones across the Midlands (fig. 10.7), so that at the end of the period, as deltas also started coming from mountains rapidly rising to the south, some of the floods seem to have covered up the coal swamps with mountain debris.

The effect of the collision of plates to the south blundered into Britain,

squeezing all the thick sedimentary
basins in the south west, and causing
granites to form beneath the high
mountains. Further to the north, all the
old faults and folds became active once
again. These deformations mark a
rather inexact boundary with the
Permian. Britain was still travelling
north, between 10 and 20 degrees
above the equator, but for a period
found itself in the middle of a large
continental mass, as oceans had now
closed to the north-west and the south.
The lush forests and swamps were
replaced by hot desert winds from the
NE and E; we were travelling through
the north tropical desert belt.

10.6 Edinburgh in the Carboniferous (above)
10.7 Birmingham in the Carboniferous (right)

127

The Permian story is almost a rerun of the Devonian; a landscape wearing down with sands deposited in small basins between the mountains. Around Britain large cracks were beginning to form in the crust, marking the newest stage of our geological history. Volcanoes appeared at several places around the North Sea. (The Whin Sill had been intruded into North England as the period began.) The first rifting in the North Sea caused sinking that led to the development of a large inland sea, the Zechstein Sea, whose waters covered much of Northern England (fig. 10.8). Periodically, the sea would retreat as it evaporated, leaving a vast area covered in salt, only to fill up again, probably from the north, along cracks associated with the rift. Sometimes sand-dunes would move in off the desert and cover up the former sea bed. To the west, much of the Irish Sea and the Inner Hebrides were valleys in which the desert sands accumulated.

By the Triassic period, the climate was becoming a little wetter, but still with long dry periods. The Worcester graben, a major valley formed by faulting, carried pebbles from mountains in France down into the Midlands (fig. 10.9). There was not only more water around, with lakes and rivers, but also less British mountain. The country was starting to take on elements of the shape we recognise today. The Irish Sea was almost entirely formed, as a trough of deposition. Sands, mudstones and salt deposits appeared in the lakes and seas, and the old landscape sank beneath the sediments. The pre-Cambrian mountains at Charnwood Forest became buried. As the lagoons extended, they joined up to form a sea

10.8 York in the Permian

10.9 Worcester in the Triassic

10.10 Mid North Sea in the Jurassic

and the shape of the country that had appeared for a short period was lost again.

By the beginning of the Jurassic, the highlands had been worn down and only parts of Scotland and the Pennines, East Anglia and Cornwall were land. The gradual sinking, that had been going on since the Permian, was interrupted in the Middle Jurassic, when there was some rather localised uplift of the northern areas, probably associated with the Atlantic opening between Africa and America. Deltas, rather like those of the Carboniferous, formed around these northern land areas, flooding sand into parts of the North Sea and the Minch.

Occasional volcanoes were still forming along the rifts (fig. 10.10). In the Middle Jurassic of Devon and Somerset there are clays, known as *Fuller's Earth*, once used for cleaning the oil from wool, that were originally deposits of volcanic ashes formed from volcanoes on the edge of the young Atlantic. Throughout central England, more of the marine limestones and clays followed the shallow water interlude, while further south, similar kinds of sediments were forming in coastal lagoons, some of fresh water.

As so often with the oscillations of the country, the next major impulse came from the side that had up to that point shown stability – the west. During the first part of the Cretaceous, the Rockall bank became split away from Britain as the North Atlantic made its first tentative appearance. The land of the west was uplifted and rivers flowed across the tilted land surface depositing deltaic sediments around Sussex and Hampshire. Erosion, and a kind of crustal relaxation, seem to have lowered the land back beneath the sea once again. The Late Cretaceous was the age of the chalk.

The sea levels, half way through the Cretaceous, seem to have been very high. (Ocean crust sinks and deepens as it gets older, so it is possible that a large number of small young oceans, existing at that time, could force excess water over the low lying continental regions.) The chalk sea is extraordinary, in that until one reaches the far northern part of the North Sea rift, there is hardly any sediment mixed with the tiny calcite coccoliths (fig. 10.11). There was virtually no erosion over Britain. While dinosaurs were roaming across America, Britain lay beneath two or three hundred metres of water. There may have been some low flat islands to the west, but, as mentioned earlier, the information is missing.

Britain was now 40°N. The second stage of the North Atlantic opening was about to begin. The peaceful chalk

10.11 Reading in the Cretaceous

seas ended as the land was uplifted, beside an ocean in formation. All around the North Atlantic this early Tertiary uplift made mountains in places where there had been no orogeny for more than 400 million years. Some disturbance in the early spreading seems to have caused more than 3,000m peaks to be raised in Western Greenland; 2,400m high ones in Western Norway; and 1,300m high ones in West Scotland; all out of the old remains of mountain belts, long eroded and levelled, through uplift without folding. All our British mountains today are 'oldies' that have been reactivated. The massive volcanic province that extended from the East coast of Greenland down through the Faeroes and into the Hebrides of Scotland was probably part of the same episode (fig. 10.12).

While the volcanic fireworks and uplift took place to the north and west, the south and east accumulated more deltaic sediments. These deltas ran into a sea, in the extreme south-east, that was an extension of the North Sea and that continued into the Hampshire and Paris Basins. That sea would persist today, with no South-East England at all, if it had not been for the Alpine orogeny, that buckled the crust more than 1,000km away from the major compression zones, causing the uplift and gentle folding of the young Tertiary, Cretaceous and Jurassic rocks of the 'home counties'.

The uplift of the ice-sculpted Highlands still continues today. Otherwise the only change the country has undergone since the Tertiary is one of climate. Along the Western side, the coastline is controlled by deep fault bounded rifts and basins. To the east, the land shelves away gently into the still sinking North Sea basin.

In summary – the last five hundred million years have seen (fig. 10.16): the loss of an ocean between England and Scotland; the formation of mountains across the country followed by erosion, deserts, deltas and shallow seas; the loss of another ocean or series of oceans to the south followed by more mountains, uplift and then erosion and deserts; inland lakes becoming joined into seas; then rifting and uplift of the old mountains on the fringe of a new ocean. All while Britain steadily drifted across the globe.

The period we live in now does not look very special. But thanks to the combination of re-uplift to the West and Alpine nudge to the South East, in combination with current low Ice Age sea levels, there is more land above the

10.13 *Leeds in the Ice Age*

10.14 *English Channel in the Ice Age*

10.12 *Northern Ireland in the early Tertiary*

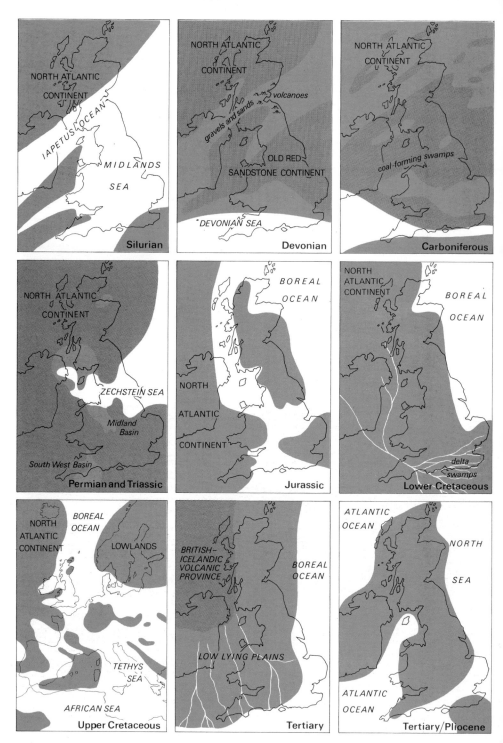

sea today than there has generally been during the geological history of the British Isles.

And what of the geology of the future? Through our understanding of the past it should be possible to make some predictions. The deltas of the Thames and the Rhine will re-advance into the North Sea; the east coast of Britain will erode fast, as in parts of Suffolk, where whole towns have vanished in the last few centuries (fig. 10.15). The sea level will rise, the mountains will wear down; the country will slowly sink beneath the waves; but with more tectonic ripples from the south, as Africa continues on its relentless journey into Europe, closing the Mediterranean and giving more uplift to Southern England.

In the more distant future the Atlantic Ocean will develop subduction zones along its margins and start to contract as the Pacific Ocean is doing today. Andesite volcanoes will form in arcs along the British Isles, till finally, as the Atlantic closes, the whole country becomes involved in another orogeny. Massive chains of mountains will form to the west and all the thick piles of sediments in the North Sea become compressed and folded. But all that is maybe 100 million years or more away (Colour 19).

Till then, we can be thankful that the turmoil recorded in the rocks of Britain is, for the moment, in abeyance and that our countryside has been made docile with age. Our scenery has mellowed from the fire, water, wind and ice that all helped to create it. But remember, beneath your feet there lies a history as complex and as turbulent, as placid and as violent as the history that we ourselves have enacted on the country's surface.

10.15 Print after Turner of the storm of 1317 which washed away much of the town of Dunwich, Suffolk, once a wealthy port, commercial centre and capital of the East Anglian area
10.16 Changes – land against sea in Britain's geological history (left)

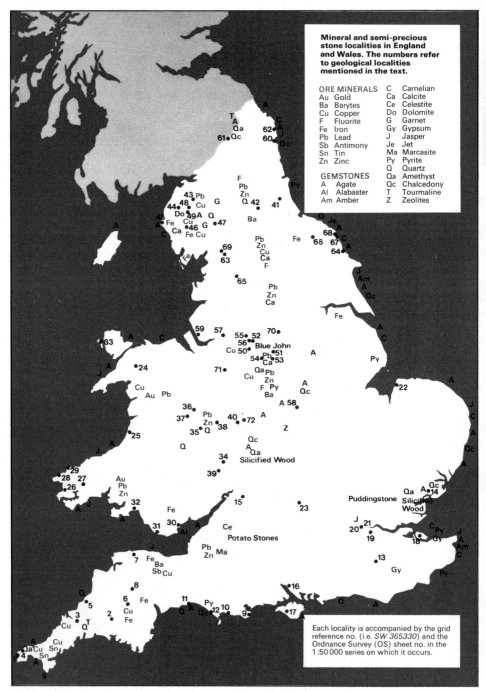

Mineral and semi-precious stone localities in England and Wales. The numbers refer to geological localities mentioned in the text.

ORE MINERALS
Au Gold
Ba Barytes
Cu Copper
F Fluorite
Fe Iron
Pb Lead
Sb Antimony
Sn Tin
Zn Zinc

GEMSTONES
A Agate
Al Alabaster
Am Amber

C Carnelian
Ca Calcite
Ce Celestite
Do Dolomite
G Garnet
Gy Gypsum
J Jasper
Je Jet
Ma Marcasite
Py Pyrite
Q Quartz
Qa Amethyst
Qc Chalcedony
T Tourmaline
Z Zeolites

Blue John

Silicified Wood

Puddingstone   Silicified Wood

Potato Stones

Each locality is accompanied by the grid reference no. (i.e. *SW 365330*) and the Ordnance Survey (OS) sheet no. in the 1:50 000 series on which it occurs.

Map of mineral and semi-precious stone localities in England and Wales. Numbers refer to geological localities in text.

SOUTH WEST ENGLAND

Geology

Towards the west of the peninsula, the dominating geological influence on both the scenery and the economy of the region, has come from the presence of a series of large granite intrusions. These have formed high-level rugged moorland scenery from Dartmoor to Land's End and the Scilly Isles. From them has come China Clay and out of them issued the fluids that caused the veins of tin and associated minerals. During intrusion the Carboniferous and Devonian sediments were domed up and metamorphosed by the granites. The folded Devonian shales and limestones outcrop in North Devon and Cornwall, while much of the rest of Devon is made up of black shales of the Carboniferous. At the Lizard and Start Point, a major thrust has forced up serpentinized peridotites, gabbros and schists. To the east of a line between Exeter and Minehead, there are thick deposits of red Permian and Triassic marls, sandstones and conglomerates. These filled up the trough that marks the separation of the South West England Carboniferous–Permian orogeny, from the rest of Britain.

Minerals

This is probably the most stimulating part of England for the mineral collector. Copper, tin and lead ores have been

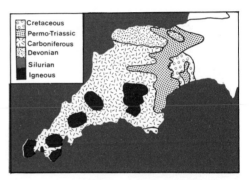

G1. *The scene around Botallack mine in Cornwall, one of Britain's premier mineral locations*

worked in Cornwall over several centuries and mineral specimens are to be found on many old mine dumps. Good areas to search include St Just, St Ives, Porthleven and Cligga Head. Gemstones, including amethyst and smoky quartz, may be located around St Austell, in the China Clay (kaolin) quarries.

North Devon was once an important mining centre, and mine dumps around Combe Martin and Minehead may bear profitable investigations today.

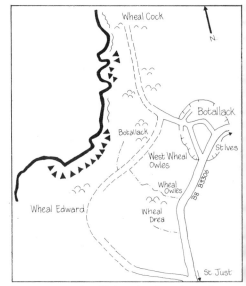

Cornwall

Botallack (1)
Nr St Just
Cornwall
SW 365330 (OS sheet no 203)

The village of Botallack is the gateway to a very beautiful part of the Cornish coast where relics of industrial archaeological interest are plentiful and mineral specimens may be obtained from old mine dumps.

A wide variety of mineral types occur on the spoil heaps associated with Botallack

G2. Tintagel – King Arthur's Castle

e, West Wheal Owles, Wheal Edward
Wheal Cock. All were tin mines, but
imens of fluorite, quartz, chalcedony,
er, copper, lead, iron and uranium
erals may be located; the uranium
erals occurring on the spoil heaps of
eal Edward.

here is an enjoyable walk around this
which takes in some of the most
uthtaking scenery in the county. The
dings of Botallack Mine stand upon a
dland called Crown's Rock.

Hill (2)
dington
rnwall
375713 (OS sheet no 201)

monumental mine chimney which
ds on the top of Kit Hill is a well
wn landmark in the district. Today the
is a popular picnic spot but old mine
dings and spoil heaps are still a feature
his once important mining area.
mples of *cassiterite*, the principal ore
n, can be located here.

tire Head (3)
Wadebridge
rnwall
924804 (OS sheet no 200)

tire Head is reached by foot from
tire. The headland is carved out of
ow lavas of Late Devonian age,
ich are extremely well exposed. Pentire
d is situated at the mouth of the river
nel on the northern shore.

thmeor Cove (4)
Zennor
rnwall
425375 (OS sheet no 203)

contact zone between granite and a
grained granite rock, known as an
te, is exposed at Porthmeor Cove.
ites are often found in association with
matites. At Porthmeor veins of *schorl*

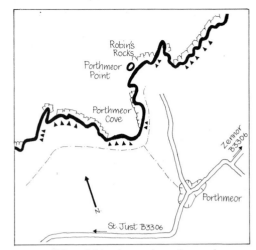

are prolific within the rock and may be
easily identified. Schorl is the black form
of tourmaline and has no decorative value.

Tintagel (5)
Cornwall
SX 060885 (OS sheet no 200)

Slates of Late Devonian age are exposed
in cliffs at Tintagel, together with a bed of
pillow lava. By the castle, quartz veins are
exposed within the rock and contain many
well developed, if rather small, crystals.

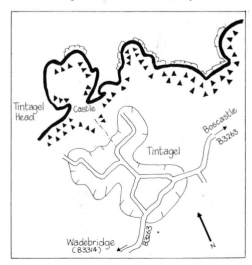

Devon

Dartmoor (6)
Devon
OS sheet no 191

The geology of selected areas of Dartmoor
is described in the Geologists' Association
Guide No 33 by W R Dearman.

Exmoor National Park (7)
North Devon
OS sheet nos 180 and 181

The Devonian sedimentary rocks of North
Devon are described with localities on
Information Sheet No 13 entitled *Exmoor
National Park 'Geology'* published by the
Exmoor National Park Authority. This is
available from the Information Centre,
Market House, The Parade, Minehead,
Somerset TA24 5NB.

Meldon Quarries (8)
Okehampton
Devon
SX 568920 (OS sheet no 191)

Permission must be obtained in writing to
visit either of the two quarries described.

The most interesting of these quarries is
the Meldon Aplite Quarry, near
Okehampton. As the name suggests, the
rock exposed within the quarry is a granite
aplite. Veins of pegmatite occur here and
specimens to be found include pink and
green tourmaline as small crystals. Quartz
crystals, axinite, and rare lithium bearing
minerals can also be found.

The second and larger quarry belongs to
British Railways Southern Region. The
rocks exposed have been contact
metamorphosed and minerals to be found
may include scheelite, idocrase and garnet.

SE AND CENTRAL ENGLAND

Geology

This is the area of younger sediments,
formed in the last 200 million years
(Mesozoic and Cenozoic). These were
buckled by the re-uplift of the older
formations to the west and by the Alpine
orogeny, a long way to the south. The
lines of hills of South East England and
scarps of the North and South Downs, I
of Wight and Chilterns are formed from
the tougher chalk resting on the softer
clays and sands of Early Cretaceous and
Jurassic Age. The chalk ridge extends int
Norfolk and up along the Yorkshire coa
The anticline across Sussex and Hampshi
has removed, through subsequent erosio
the Tertiary clays and sands that once
connected the London basin with the
Hampshire basin. Still more recent shell
and coral deposits cover large areas of
Norfolk and Suffolk. The sediments
beneath the Cretaceous are Jurassic clays

Pliocene
Oligocene & Eoce
Cretaceous
Jurassic

and limestones – the limestone forming the Cotswold hills – the softer clays are seen best in cliff sections on the Dorset coast.

Much of the area to the north of London contains recent deposits of glacial boulder clay, glacial gravels and river silt.

Minerals

Mineral occurrences are scarce in this area of Britain although gemstones are surprisingly common.

The silica gemstones which include chalcedony, carnelian, agate, jasper and amethyst may be found as pebbles in the fluvio-glacial deposits of Central and South East England. Gravel pits near Bromsgrove in Worcestershire, Lichfield in Staffordshire, Retford and Netherfield in Nottinghamshire and Colchester in Essex, will all yield interesting examples. The extreme south and south east of the area are free from Ice-Age pebble deposits as the glaciation did not proceed to the south of a line between the river Severn and the river Thames.

The beaches of Yorkshire, Lincolnshire and East Anglia also yield gemstone pebbles and erratics – rocks carried by the glaciers – some from as far away as Norway.

G3. Late Tertiary, Alpine folding in Mesozoic – sediments at Lulworth Cove

Dorset

Durlston Bay (9)
Nr Swanage
Dorset
SZ 035780 (OS sheet no 195)

Durlston Bay is reached via Swanage and is situated immediately to the south of the town. The Purbeck Beds which form the top of the Jurassic succession in Britain are exposed here and a variety of fossils may be located on the beach, including fish scales and turtle bones.

This locality is of special interest for it is one of very few sites in the world where the bones of primitive Jurassic mammals have been found. During the age of dinosaurs, these, our mammal ancestors were smaller than cats. The fossils were found during excavations for a building project on the cliff top.

Lulworth Cove (10)
Dorset
SY 826798 (OS sheet no 194)

Of particular interest are the exposures of the Fossil Forest which is visible approximately 1km to the east of the cove. The fossil tree trunks exposed here are of the Lower Purbeck beds of Jurassic Age and are revealed near the cliff top. Beneath the trees is a dirt bed which is said to be the soil in which the trees grew.

Lyme Regis–Charmouth (11)
Dorset
SY 341921 (OS sheet no 193)

The Lyme Regis–Charmouth area is famous for the rocks of the Lower Jurassic (Lias) which form cliffs in the area. Unfortunately the cliffs are very unstable and are therefore highly dangerous. Specimens of local fossils may be found upon the beach, which removes the necessity to explore the cliff faces.

Examples of several minerals and gemstones can be found as pebbles on the beaches here. These include pyrite, selenite, chalcedony, quartz crystals and the occasional specimen of agate.

The geology of the area is covered by the Geologists' Association Guide No 23 by D V Ager and W E Smith. This and all of The Geologists' Association Guides are available from 'The Scientific Anglian' 30/30A St Benedicts Street, Norwich NOR 24J.

Poole–Chesil Beach (12)
Dorset
OS sheet no 194

The geology of the Dorset coast between Poole and the extraordinary Chesil Beach is described by M R House in the Geologists' Association Guide No 22.

East Sussex–Kent

The Weald (13)
OS sheet nos 187 and 188

The sedimentary geology of the Weald is described in two Geologists' Association Guides. J F Kirkaldy outlines the geology of the Weald in Guide No 29 while the Hastings Beds of the central Weald are discussed in Guide No 24 by P Allen.

Essex

Colchester (14)
Essex
OS sheet no 168

Of particular interest in the Colchester area are the numerous gravel pits which are working deposits dropped by rivers during the Ice Age. Flint pebbles make up the bulk of the gravel but specimens of chalcedony, carnelian, agate, amethyst, and silicified wood can be found with reasonable ease. Fossils are also plentiful particularly sea urchins (echinoids) which are preserved in flint. Permission to enter the gravel pits must be obtained from the respective pits.

G4. Chalk scarp of the North Downs

Gloucestershire

Cotswold Hills (15)
Gloucestershire
OS sheet no 172

The geology of the Cotswold Hills is described in the Geologists' Association Guide No 36 by D V Ager and D T Donovan.

Hampshire–West Sussex

Southampton (16)
OS sheet no 126

The geology of several coastal sections in Hampshire and Sussex near Southampton is described in the Geologists' Association Guide No 14 by D Curry and D E Wisden.

Isle of Wight

Isle of Wight (17)
OS sheet no 196

The Cretaceous and Tertiary geology of the Isle of Wight is described in the Geologists' Association Guide No 25 by C W Wright and D Curry.

G5. Landslipping at Alum Bay, Isle of Wight

Kent

Warden Point (18)
Isle of Sheppey
Kent
TR 020725 (OS sheet no 178)

At Warden Point the London Clay is exposed as cliffs which lie behind an interesting shingle beach. Concretions are quite common on the beach and may contain green calcite and small rosettes of barite crystals. Colourless gypsum, in the form of selenite, is particularly common and although many of the crystals are waterworn, good examples, including twinned crystals, may occur.

The cliffs at Warden Point are very unstable, particularly in wet weather and care must therefore be taken. Access to the beach here can be obtained by detouring around the perimeter of the caravan site.

London

Charlton Sand Pit (19)
Greenwich
London
OS sheet no 177

This sand pit lies within Maryon Park, but access is only possible by permit which is obtainable from the Borough Engineer, London Borough of Greenwich, Churchill House, Greens End, Woolwich, London SE18.

The pit is the type locality for the Woolwich Beds and offers a complete section through the Lower Tertiary sequence.

London (20)
OS sheet nos 176 and 177

The geology of the London Region is described in the Geologists' Association Guides No 30 North of the Thames, and No 30b South of the Thames.

London (Geological Walks) (21)

A series of geological walks in London have been prepared by the Geological Museum.

The walks identify and outline the background to rocks which have been exploited for building purposes in the London Area. Small guides are available from the Geological Museum, Exhibition Road, South Kensington, London. Titles already available include 'Shopfronts in Bond Street', and 'Stone in South Kensington'. Future guides include 'Stone in the West End', 'Westminster Rock', 'Stone in the City' and 'Stone on the South Bank'.

Norfolk

Hunstanton (22)
Norfolk
TF 675410 (OS sheet no 132)

The cliffs immediately to the north east of Hunstanton are composed of rocks of Cretaceous Age. Red chalk occurs here with Lower Greensand and examples of both can be examined on the beach. Specimens of such fossils as belemnites, brachiopods and sea urchins may be found on the beach having been eroded from the cliffs by the sea.

Oxfordshire

Oxford (23)
OS sheet no 164

The geology of the area around Oxford is described in the Geologists' Association Guide No 3 by W S McKerrow.

'ALES AND BORDERLANDS

eology

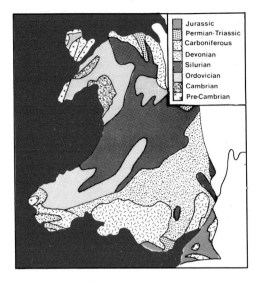

s we approach Wales, the mountains gin, hewn out of the thick piles of diment of Cambrian, Ordovician and lurian Age, and folded during the aledonian orogeny. Pieces of the pre-ambrian basement rocks below these diments form the Malvern Hills, the ngmynd in Shropshire and almost the hole of Anglesey. In general, the thick dimentary trough moved towards the uth east, so that the Cambrian rocks of ales and sandstones are thickest to the rth west, around the mountains of the arlech Dome. The Ordovician volcanic hes, lavas, shales and greywackes make much of the region around Snowdon – so further to the east near Oswestry. The lurian rocks consist of shallow water nestone formations in the borderlands

with a thick pile of deeper water sediments towards Central Wales. The greatest amounts of re-uplift are in areas to the north-west where intrusions associated with

6. Folded Palaeozoic – Cardigan coast near Aberystwyth

the Ordovician volcanoes underlie Snowdon and Cader Idris.

To the south east of Wales there are thick deposits of Devonian red sandstones and marls that make up the Brecon Beacons. In some areas these pass directly into the Carboniferous limestones that outcrop at a number of areas of South Wales and round into Somerset. Above these, around Bristol, the Forest of Dean and South Wales, there are the deltaic sediments of the Coal Measures. In all these areas, particularly in South Wales, these sediments were folded by compression from the Hercynian orogeny to the south.

Minerals

Many of the beaches of Wales will provide small pebbles of agate, chalcedony and jasper. Beaches near Tenby in Dyfed and on the Lleyn Peninsula contain reasonable specimens.

Lead and zinc ores have been mined in several areas in Central and North Wales, and mine dumps around Minerva and Pumpsaint are worthwhile hunting grounds. Gold can be panned in small amounts from streams near Dolgellau.

Pary's Mountain on Anglesey was the scene of open cast copper mining in the past and examples of chalcopyrites and malachite may be located.

On the Welsh Border, in Shropshire the main area for minerals is south of Shrewsbury near the village of Shelve. The principle ore minerals to be found here are galena and sphalerite but quartz is common together with a number of secondary ores.

G7. Boulder clay cliffs lie behind this beach on the Lleyn Peninsula, Caernarvon. Beaches in the area have pebbles of semi-precious stones

Somerset has the famous 'potato stones' (geodes) to offer. These are usually found around the fringes of the Mendip Hills in association with a red Keuper (Triassic) marl which overlies the Carboniferous Limestone. The geodes may contain crystals of milky quartz, smoky quartz or amethyst, together with small needles of goethite.

Caernarvon

Snowdonia (24)
Caernarvon
OS sheet no 115

The complex igneous geology of Snowdonia is described in the Geologists' Association Guide No 28 by D Williams and J G Ramsay.

Dyfed Region: Cardigan

Aberystwyth (25)
Cardigan
SN 580818 (OS sheet no 135)

Exposures of Silurian grits, of which Aberystwyth is the type locality, are well exposed on the coast to the north and south of the town. The turbidite beds are interbedded with mudstones and exhibit a number of interesting sedimentary features including current and graded bedding.

Dyfed Region: Pembrokeshire

Little Haven (26)
St Brides Bay
Pembrokeshire
SM 855130 (OS sheet no 157)

Shales and sandstones of the coal measures are well exposed in the cliffs at Little Haven. The Carboniferous rocks exhibit strong folding and faults of Hercynian Age. (See map at bottom of p. 145)

Pembrokeshire (27)
OS sheet nos 145, 157 and 158

The geology of Pembrokeshire is interestingly described, with localities, in the small guide 'The Rocks' by Dr B S John, which is part of the Pembrokeshire Handbook Series. Copies are available from E A Roberts, Y Garn, Llanychaer.

Pembrokeshire Coast (28)
OS sheet nos 157 and 158

The Pembrokeshire Coast National Park Committee offer a programme of guided walks, accompanied walks and lectures on the geology and natural history of this attractive coastline. Further details including guides, are available from the Pembrokeshire Coast National Park Information Service, County Office, Haverfordwest, Dyfed.

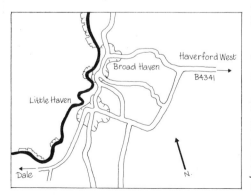

G9. Lower Palaeozoic slates – Snowdon from 'Castle of the Winds', Caernarvonshire

Strumble Head (29)
Pembrokeshire
SM 892414 (OS sheet no 157)

Pillow lavas of Ordovician Age are exposed to the north of the lighthouse on Strumble Head. The lavas can be reached by the path which runs to the right of the lighthouse.

Glamorgan Region

Cardiff (30)
South Glamorgan
OS sheet no 171

The geology of the Cardiff area is described in the Geologists' Association Guide No 16 by J G C Anderson.

The Glamorgan Coast (31)
OS sheet nos 159 and 170

A guide to the geology of this area entitled 'The Glamorgan Heritage Coast', published by the University College, Cardiff is now available. This coastline offers an interesting array of Carboniferous, Triassic and Jurassic rocks, faults, and many features of coastal erosion.

Swansea (32)
West Glamorgan
OS sheet no 159

The geology of the Swansea area is described by T R Owen, and F H T Rhodes in the Geologists' Association Guide No 17.

Gwynedd Region: Anglesey

Holyhead (33)
Anglesey
OS sheet no 114

Green schists of the Mona Complex are exposed to good effect in Holyhead Harbour. These metamorphosed sedimentary rocks, of pre-Cambrian Age, are intensely folded and provide an interesting insight into the power of deformation and orogeny.

Hereford and Worcester

Malvern Hills (34)
OS sheet no 162

The complex geology of the Malvern Hills is described in the Geologists' Association Guide No 4.

Salop

Mortimer Forest (35)
Ludlow
Salop
SO 510750 (OS sheet nos 137 and 138)

This nature trail is based on the Ludlow series of sedimentary rocks of Silurian Age, which include the Wenlock Limestone. The geology of the trail is described in the guide to the *Mortimer Forest Geological Trail* published by the Nature Conservancy Council obtainable from The Regional Officer, Attingham Park, Shrewsbury, Shropshire S74 4TW.

Pontesford Hill (36)
Pontesbury
Salop
SJ 410050 (OS sheet no 126)

Pontesford Hill is formed out of Uriconian, pre-Cambrian, volcanic rocks which have been intruded by a large mass of basaltic material (an olivine dolerite). Habberley Brook flows down a wooded valley on the eastern side of the hill in which the Pontesford shales of Ordovician Age are exposed. The hill is situated

G8. *Pre-Cambrian Mona complex – South Stack, Holyhead in Anglesey*

imediately to the south east of
ontesbury on the A488 Shrewsbury–
ishop's Castle road.

nelve–Snailbeach (37)
alop
O 336990 (OS sheet no 137)

nce an important mining centre, the
ountryside around the villages of Shelve
nd Snailbeach is rich in mining remains.
ld mine dumps are a common sight and
variety of minerals may be found. Likely
nds include sphalerite (zinc sulphide)
alena (lead sulphide) and quartz.

outh Shropshire (Salop) (38)
OS sheet no 137

number of geological itineraries in South
hropshire are described by W T Dean in
e Geologists' Association Guide No 27.

Velsh Borderland

Velsh Borderland (39)
OS sheet nos 126, 137 and 149

he geology of the Silurian inliers of the
Velsh Borderland (south-eastern) is
escribed in the Geologists' Association
uide No 5.

Vest Midlands

Vren's Nest Hill (40)
Dudley
O 936917 (OS sheet no 139)

ilurian rocks including the Wenlock
imestone and Wenlock Shale are exposed
ere in old quarry workings.
   The geology of the Hill is described in
e guide to the Wren's Nest National
Nature Reserve Geological Trail which is
ublished by the Nature Conservancy. It is
btainable from the NCC, 19 Belgrave
quare, London SW1X 8PY.

MIDLANDS AND N ENGLAND

Geology

Although containing rocks from the pre-
Cambrian through to the Permian and
Triassic the dominating influence on the
general scenery and economics has come
from rocks of the Carboniferous.
However, as if attached on to this area,
there is the Lake District that contains an
older series of rocks similar to those of
North Wales. The mountains of Cumbria
as well as those of the Isle of Man are the
result of the recent uplift of large
underlying granite intrusions. In the Isle of
Man there is a thick pile of Cambrian
shales and grits with, at either end of the
island, Carboniferous limestones and, at the
north end, Permian-Triassic red sands. The
Lake District is a tilted thick pile of shales
and sandstones of Ordovician Age (the
Skiddaw Slates) with above, and to the
south, about 3,000m of andesite and

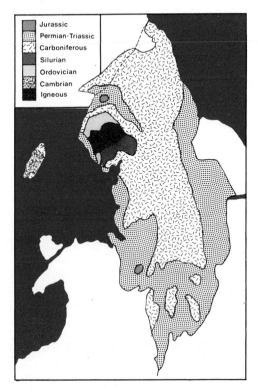

rhyolite ashes and lava flows. Minor amounts of Silurian rocks occur to the south – the whole area has suffered doming and recent glacial erosion.

Other areas of pre-Carboniferous rocks are relatively small. To the south, in the Midlands, the pre-Cambrian basement peeps through the sedimentary cover around Charnwood and Nuneaton. Other areas of basement rocks are seen around Cross Fell where, beneath the Carboniferous, a similar series of rocks to those of the Lake District, is repeated and at Cow Green in Teesdale where Ordovician rocks outcrop.

The Carboniferous rocks lie in a broad anticline across the Pennines and Derbyshire Dales with the older Carboniferous limestones in the middle and the younger Millstone Grit and Coal Measures formations lying to either side. Permian and Triassic red sediments and evaporites lie in basins around the Lake District and to the south and east of the Carboniferous rocks that form the spine of the North of England.

Igneous rocks are generally scarce except for the volcanic rocks in Derbyshire of Carboniferous Age and the Whin Sill of the Carboniferous–Permian boundary that punctuates the scenery at a number of cliffs and crags and the Devonian volcanics of the Cheviot Hills.

Minerals

The Pennine Hills have been a rich source of minerals during the last three centuries. Lead mining has gone on here since Roman times, and old mine dumps and mining relics are common from Derbyshire to Northumberland.

Good centres from which to search for minerals are Castleton and Matlock in Derbyshire; Grassington and Reeth in Yorkshire; Stanhope in Co. Durham; Alston in Cumbria and Blanchland in Northumberland. Specimens of lead and zinc ores together with barite (barium sulphate), fluorite, quartz and calcite are

common throughout the Pennines. Localities for copper minerals are less abundant.

The Lake District contains a wide variety of mineral types, but being one of the most popular areas of Britain, many of the sites have been well picked over by previous generations of collectors. Good sites for minerals include Caldbeck Fells and Cleator Moor.

The volcanic complex of the Cheviot Hills contains a range of silica gemstones which are derived from the volcanic rocks. Specimens of agate, chalcedony and carnelian may be picked up amongst the shingle in all the streams and rivers which cut through the volcanic rocks. Good centres from which to explore the Cheviot are Rothbury and Wooler. Small crystals of tourmaline may be found on quartz originating in the crush zones of the Cheviot granite which lies in the centre of the volcanic assemblage.

G10. Cockscombe barite from Bonsall Moor, Derbyshire

Co Durham

Durham (41)
OS sheet no 88

The geology of the countryside around the City of Durham is described in the Geologists' Association Guide No 15 by G A L Johnson.

148

High Force (42)
Nr Middleton in Teesdale
Co Durham
NY 880284 (OS sheet no 92

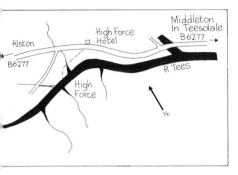

The well known intrusion called the Whin
Sill is extensively exposed near Middleton
In Teesdale where its columnar structure
dominates the landscape. The sill also
provides the lip over which the river Tees
plunges at the famous waterfall, High
Force.

At High Force the sill shows well
developed columnar jointing which is
interspersed with two thin beds of shale.
At the base of the sill, shale, siltstone and
limestone may be observed.

It is considered that the intrusion is of
late Carboniferous or early Permian. The
huge volume of magma must have taken a
considerable length of time to be fully
emplaced.

Cumbria

Dry Gill (43)
Caldbeck
Cumbria
NY 322345 (OS sheet no 90)

This location in the Ordovician (Drygill)
Shales, is on the east side of Carrock Fell
and is reached by leaving the A66.
Penrith–Keswick road to take the
unclassified road to Mungrisdale. Continue
through Mosedale for 2.5km, bearing left
at the road junction. Leave cars by the ford

at Carrock Beck and continue along the
rough track which goes off to the left.
When the track divides after approximately
2km, bear left to Dry Gill.

The spoil heaps exposed at Dry Gill are
black in colour due to the presence of
manganese, although there is a large
amount of quartz, including some fine
crystals. This location is famous for the
occurrence of the mineral campylite which
is a form of mimetite, a lead chloro-
arsenate. Crystals of this mineral may be
found here quite easily and are golden
brown in colour. Yellow crystals of
mimetite and green crystals of
pyromorphite may also be found, usually
on quartz.

On the return journey a detour can be
made to the Driggith Mine where more
examples of the lead minerals can be found
on the mine dumps.

Embleton (44)
Nr Cockermouth
Cumbria
NY 172305 (OS sheet no 89)

Near Embleton the Skiddaw Slates (shale,
grits and sandstone) have been penetrated
by minor igneous intrusions of dioritic
lamprophyres. The intrusions can be seen
in exposures around the A66 near
Embleton; on Slate Fell, by the Golf
Course in a small quarry; and at Close
Quarry which is reached by taking the
small lane by the Wheatsheaf.

G11. Columnar jointing in the Whin Sill at
Teesdale, Co. Durham

Fleswick Bay (45)
St Bees Head
Cumbria
NX 946133 (OS sheet no 89)

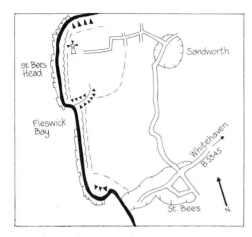

St Bees Head may be approached on foot from the village of St Bees to the south, or along the lighthouse road from the east.

Fleswick Bay is backed by cliffs of the red 'St Bees' sandstone which is of Permian Age. Jointing in the sandstone is visible, and on the north side of the Bay the joints have been opened out into small caves by erosion. The same erosion by wave carried beach material has resulted in pronounced undercutting at the base of the sandstone cliffs.

The abundant beach pebbles are mainly derived from glacial deposits close by. Most of the pebbles are of Lake District origin but include examples of such semi-precious stones as common chalcedony, agate, carnelian and jasper, which is usually red in colour.

The Lake District (46)
OS sheet nos 90 and 97

The geology of the Lake District is described in the Geologists' Association Guide No 2 by G H Mitchell.

Shap (47)
Cumbria
NY 554084 (OS sheet no 90)

The famous porphyritic granite is worked in a quarry 7km to the south of Shap. Permission to visit this and the blue quarry nearby can be obtained in writing from the main works office. Collecting is allowed.

The granite is worked for its ornamental value and mineral specimens are not common in the quarry although pyrite, molybdenite, and more predictably, quartz and feldspar crystals have been reported.

Shap Blue Quarry is working in Borrowdale Volcanics which have undergone metamorphism due to the close proximity of the granite. Minerals here include grossular and andradite garnet, epidote, and haematite.

Sinen Gill (48)
Lonscale Fell
Skiddaw
Cumbria
NY 300283 (OS sheet no 90)

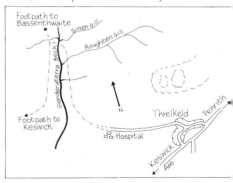

Access is by way of the village of Threlkeld and cars may be parked by the Sanatorium. From the hospital a path runs parallel with the Glenderaterra Beck up the valley.

By examining exposures along the beck the blue slates of Skiddaw can be seen to have undergone metamorphism which becomes increasingly intensive as one proceeds uphill. Reaching Sinen Gill, the

G12. The Blue Quarry at Shap in Cumbria

Skiddaw Granite can be seen at the prominent waterfall. Further up the gill a band of white quartz is exposed above the granite which in its turn is capped by a cordierite *Hornfels* (a contact metamorphosed rock). Kaolinisation of the granite is evident below the waterfall. Spoil heaps associated with the old Glenderaterra Mine contain specimens of galena, sphalerite, pyromorphite, barite and quartz.

Walla Crag (49)
Nr Keswick
Cumbria
NY 275215 (OS sheet no 90)

Walla Crag is formed out of volcanic rocks which belong to the Borrowdale Volcanic Series. Cat Ghyll flows down off the crag and has exposed sections through lava and ash.

The lavas tend to be highly vesicular and have quartz, chalcedony, agate, carnelian and jasper occurring as amygdales. Specimens of these semi-precious stones and also epidote may be found here on a scree, at the top of Cat Ghyll.

Derbyshire

Calton Hill Quarry (50)
Nr Buxton
Derbyshire
SK 117715 (OS sheet no 119)

This disused quarry on the A6, 8km east of Buxton, was opened on the site of a volcanic vent of Early Carboniferous Age. Although much of the quarry has now been landscaped, the centre of the site close to the vent is being preserved.

A platform of columnar basalt can be observed in the quarry and evidence of hydrothermal activity is to be found in the form of the minerals quartz and calcite, which occur as amygdales and in vugs within the igneous rock. This site was once popular with mineral collectors due to the presence of amethyst and smoky quartz crystals, but collecting is no longer allowed. Of special interest to geologists are the basaltic xenoliths containing olivine pyroxene and spinel, that were brought up from the mantle.

Applications to enter the site should be sent to Estates Department, Derbyshire County Council, County Offices, Matlock.

Duckmanton (51)
Nr Chesterfield
Derbyshire
SK 442708 (OS sheet no 120)

Duckmanton Railway Cutting, now disused, contains exceptional exposures within the Coal Measures. The Clay Cross Marine Band is exposed at the stage boundary between the Westphalian A and B stages of the Carboniferous. Several outcrops of non-marine shale occur within the cutting together with a number of coal seams.

Permission to visit the cutting must be sought in advance from The Conservation Officer, Derbyshire Naturalists Trust, Estates Office, Twyford, Barrow on Trent, Derby.

G13. Dovedale in the Peak District near Buxton

Mam Tor (52)
Nr Castleton
Derbyshire
SK 128835 (OS sheet no 110)

Mam Tor is one of the most scenic geological locations in the Pennines. The exposed face is what geologists call a *landslip backscar*; the front of the hill having slipped downwards leaving the scar as a cliff behind.

Subsidence, on a large scale, over many years, has earned Mam Tor the name the 'Shivering Mountain'. The Sheffield–Chapel-en-le-Frith road, the A625, that runs over the landslip has been closed on many occasions due to repeated subsidence.

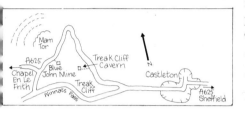

Matlock (53)
Derbyshire
SK 310600 (OS sheet no 119)

A motorised trail of geological and industrial archaeological interest in the Matlock area is detailed in Guide No 3 'Lead Mining, published by the Arkwright Society, Tawney House, Matlock, Derbyshire.

Mill Close Mine (54)
Nr Darley Dale
Derbyshire
SK 255615 (OS sheet no 119)

This location is found by turning off the A6 in Darley Dale and driving through Darley Bridge before turning off to the left into Clough Wood. A short walk into the wood leads to the ruined engine house on Watt's Shaft which was part of the old Mill Close Mine.

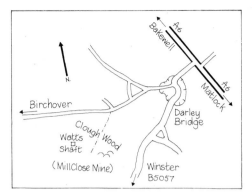

The Mill Close Mine was Derbyshire's most successful lead mine, only closing in 1939. Spoil heaps occur around Watt's Shaft and specimens of calcite, fluorite, pyrite and marcasite are reasonably common.

The Peak District (55)

The Peak Park Planning Board has published guides to a number of nature trails in the area which include details of local geology. The guides available include Padley Gorge, Tideswell Dale, Lathkill Dale, and Edale; and are obtainable from the Planning Board at Aldern House, Bakewell, Derbyshire.

Treak Cliff (56)
Castleton
Derbyshire
SK 135828 (OS sheet no 110)

Treak Cliff is the home of Derbyshire's unique form of fluorite called 'Blue John'. Fourteen different veins of Blue John occur in this hill and they have been worked from three mines. Two of these mines, the Treak Cliff Cavern and the Blue John Mine, are now open to the public and the decorative stone can be seen in situ within the Carboniferous Limestone. The third mine which falls on National Trust property is called the Old Tor Mine and is situated in the Winats Pass, on the south east of Treak Cliff. Worked examples of the stone can be purchased locally.

Greater Manchester

Manchester (57)
OS sheet no 109

The geology of the Manchester area is described in the Geologists' Association Guide No 7 by R M C Eager, F M Broadhurst and J Wilfred Jackson.

Leicestershire

Beacon Hill (58)
Charnwood Forest
Leicestershire
SK 510149 (OS sheet no 129)

The crags of Beacon Hill are composed of fine grained silicified ashes which exhibit well defined banding. These rocks, which represent a pre-Cambrian period of vulcanism, were rocky hills during Permian and Triassic times, but were eventually buried beneath Triassic sediments. Subsequent erosion has removed some of the Triassic rocks and the pre-Cambrian rocks stand out as hills again.

From Beacon Hill excellent views of the Soar Valley can be obtained; the valley floor being composed of Triassic rocks and alluvium.

Merseyside

Liverpool (59)
OS sheet no 108

The geology of the Liverpool area is described in the Geologists' Association Guide No 6.

Northumberland

Amble (60)
Druridge Bay
Northumberland
NU 265045 (OS sheet no 81)

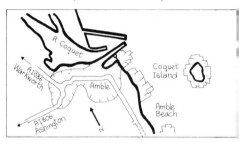

The beach at Amble offers an interesting selection of pebbles including semi-precious stones which originated in the Cheviot Hills. Agate and carnelian are quite common here although specimens are usually small. Amethyst may be found, usually as the crystalline centre of an agate.

Specimens of fossil root also occur among the shingle at Amble and a fossil forest is exposed at low tide on the beach at Druridge Bay. All the beaches here are backed by sand dunes which overlie rocks of the Coal Measures.

Cheviot Hills (61)
OS sheet nos 74 and 80

The geology of the Devonian volcanic complex of the Cheviot Hills is described in the Geologists' Association Guide No 37 by S I Tomkeieff.

Cullernose Point (62)
Howick
Northumberland
NU 261188 (OS sheet no 81)

The Whin Sill, a dolerite intrusion, is exposed in the cliff at Cullernose Point, between Alnmouth and Embleton. The sill is well exposed, lying above folded Carboniferous Limestone. Cullernose Point can be reached from the unclassified road near Howick.

North Yorkshire

Clapham (63)
North Yorkshire
SD 745692 (OS sheet no 98)

Exposures of Carboniferous Limestone and associated plant life are described in the Reginald Farrer Nature Trail at Clapham. A guide to the trail is available from the Yorkshire Dales National Park Committee, Yorebridge House, Bainbridge, Leyburn, North Yorkshire.

Cornelian Bay (64)
Nr Scarborough
North Yorkshire
TA 061560 (OS sheet no 101)

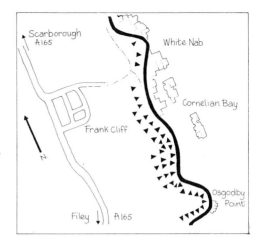

The Cliffs at Cornelian Bay are composed of glacial deposits which contain semi-precious stones. The beach below the boulder clay cliffs will yield examples of several semi-precious gemstones including common (grey) chalcedony, agate and carnelian. Specimens are usually small, but their translucent character is a considerable aid to identification, especially in strong sunlight.

Access to this beach is either across the beach from South Bay, Scarborough at low tide, or by a path which leads down the cliff from Cornelian Drive, on the A165 Scarborough–Filey Road.

G14. *The Carboniferous limestone scar at Malham with Malham tarn in centre*

Malham (65)
North Yorkshire
SD 901630 (OS sheet no 98)

The geology of the fascinating limestone scenery around Malham is described in a booklet entitled 'The Geology of the Yorkshire Dales' by A A Wilson. This guide is available from the Yorkshire Dales National Park Dept, 'Colvend', Hebden Road, Grassington, Skipton, N. Yorkshire.

North Yorkshire Moors (66)
OS sheet no 100

Several National Park Forest Trails have been established by the Forestry Commission on the North Yorkshire Moors. Details of the local geology are included where appropriate. The trails are covered by a number of guides which may be obtained from The Forestry Commission, 42 Eastgate, Pickering, North Yorkshire. The guides available include the Silpho, Wykeham, Sneverdale, Newtondale, and Falling Foss Trails.

Ravenscar (67)
North Yorkshire
NS 980018 (OS sheet no 94)

Excellent exposures of Triassic rocks, which include alumshale and ironstone, Jurassic rocks and Glacial Deposits can be examined in this area. Details may be found in the guide to the Ravenscar Geological Trail, which can be obtained from Information Service, The Old Vicarage, Bondgate, Hemsley, North Yorkshire YO 65BP.

G15. Much coastal erosion has taken place since this old photograph of Robin Hood's Bay, Yorkshire was taken

Robin Hood's Bay (68)
North Yorkshire
NZ 958035 (OS sheet no 94)

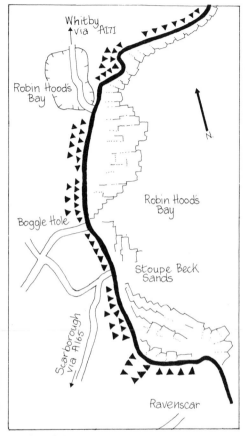

Whitby
↑ via A171

Robin Hoods
Bay

N.

Robin Hood's
Bay

Boggle Hole

Stoupe Beck
Sands

Scarborough
via A165

Ravenscar

Rocks of the Lias (Jurassic) are exposed as cliffs near the village of Robin Hood's Bay. The rocks also form the wave-cut platform which is exposed at low tide. Glacial deposits in the form of boulder clay overlie the Lias and become the dominant feature in the cliffs near Stoupe Beck.

Pebbles of porphyry of Scandinavian origin and the unusual Shap Granite occur on this beach and provide evidence of ice movement to the area, at different times, from Scandinavia and the Lake District. Semi-precious stones, also found amongst the shingle, are of Cheviot or Scottish origin. Access to the beach is from Robin Hood's Bay, Boggle Hole or Ravenscar.

White Scar Caves (69)
Ingleton
North Yorkshire
SD 730683 (OS sheet no 98)

White Scar Caves are situated in the Great Scar Limestone and are open to the public most of the year round. Typical of cave systems in limestone country, the caves contain calcite in the form of stalactites and stalagmites as well as 'flow stone' which can be seen on the walls of the caves.

South Yorkshire

Sheffield (70)
South Yorkshire
OS sheet no 111

The geology of the Sheffield area is described in the Geologists' Association Guide No 9 by C Downie.

Staffordshire

Stoke-on-Trent (71)
Staffordshire
OS sheet no 118

The geology of the area around Stoke-on-Trent is described in the Geologists' Association Guide No 8 by F Wolverson Cope.

West Midlands

Birmingham (72)
OS sheet no 139

The geology of the Birmingham area is described in the Geologists' Association Guide No 1 by P A Garrett, W G Hardie, J D Lawson and F W Shotton.

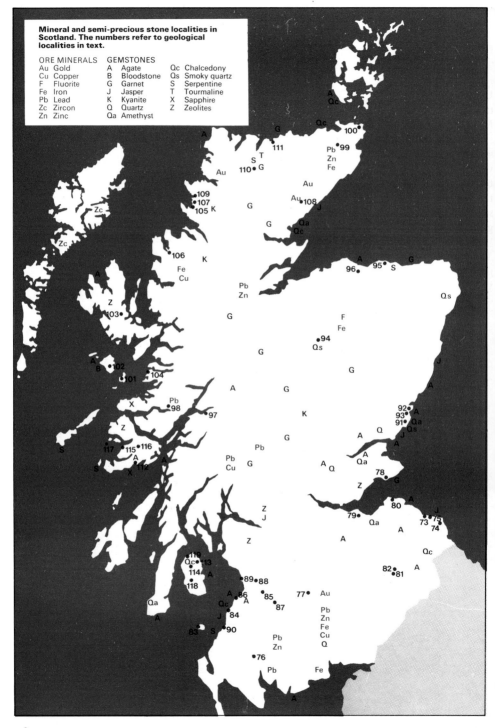

Mineral and semi-precious stone localities in Scotland. The numbers refer to geological localities in text.

ORE MINERALS
Au Gold
Cu Copper
F Fluorite
Fe Iron
Pb Lead
Zc Zircon
Zn Zinc

GEMSTONES
A Agate
B Bloodstone
G Garnet
J Jasper
K Kyanite
Q Quartz
Qa Amethyst

Qc Chalcedony
Qs Smoky quartz
S Serpentine
T Tourmaline
X Sapphire
Z Zeolites

SOUTH SCOTLAND

Geology

The area contains two distinct belts of rocks: the mountains of the Southern Uplands extending into the Cheviots; and the relative lowlands of the Scottish Midland Valley.

The Southern Uplands are made up of steeply dipping thick sedimentary piles of Ordovician and Silurian age, that formed to the south of mountains that during those periods were rising to the north. The sediments are a mixture of shales and greywackes with some basaltic pillow lavas and serpentinites – as at Girvan. Within these uplands there are a few large granite intrusions – all about 400 million years old.

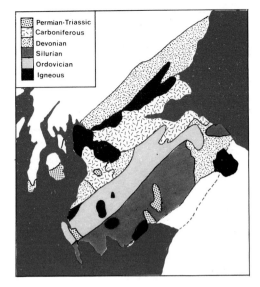

Permian-Triassic
Carboniferous
Devonian
Silurian
Ordovician
Igneous

G16. *Agate pebbles as found on the east coast of Britain*

The Midland Valley is a collapsed crustal block – almost a rift valley, lying between the lesser mountain province to the south and the Highlands to the north. Set in a kind of broad syncline, Devonian Old Red Sandstones lie to either side and include basaltic andesite volcanics. The thick Devonian is followed by Carboniferous lagoonal and deltaic sediments – seen in the central parts of the valley. During this period there were many active volcanoes in the region and some of the isolated hills are eroded remnants of volcanic necks.

Minerals

Of special interest are the nickel ores which may be located, with difficulty, on spoil heaps near Newton Stewart, and stibnite the antimony ore which has been extracted at 'The Knipe' near New Cumnock, and in Eskdale.

Many minerals may be found around the village of Leadhills in Lanarkshire which was once an important mining centre. Lead, zinc, copper and iron minerals are especially common here.

The large granite masses which form Cairnsmore of Fleet and Merrick in

Kircudbright have been credited with beryl, but it is doubtful if crystals of any size occur today.

Agate, the silica gemstone, is common within the Midland Valley, among the volcanic rocks of the Ochil, Sidlaw and Pentland Hills. Specimens may be found in ploughed fields during the winter months. Beaches near Montrose in Angus, Tayport in Fife, and Dunure in Ayrshire also yield pebbles of agate, which used to be called 'Scotch Pebbles'.

Zeolite minerals and jasper occur in the volcanic rocks of the Campsie Fells north of Glasgow and in the Renfrewshire Uplands.

Borders Region: Berwickshire

Cove Harbour (73)
Nr Cocksburnpath
Berwickshire
NT 787718 (OS sheet no 67)

One mile north of Siccar Point, Carboniferous rocks are exposed along the shore.

Calciferous Sandstones make up the bulk of the exposure and examples of plant fossils, brachiopods, gastropods and crinoids can be seen in situ.

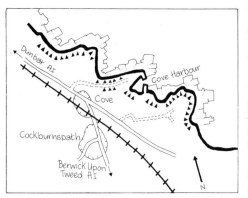

Pettico Wick (74)
Nr St Abb's Head
Berwickshire
NT 907692 (OS sheet no 67)

Greywackes and shales of Silurian Age are exposed in the cliffs here in the form of a broad horizontal syncline. These Silurian rocks are separated from the Lower Old Red Sandstone lava, which forms much of St Abb's Head, by a fault which crosses the bay. The folded Silurian rocks can be viewed from St Abb's Head.

Siccar Point (75)
Nr Cocksburnpath
Berwickshire
NT 812709 (OS sheet no 67)

Siccar Point is situated to the north of St Abb's Head and is reached from the A1107. The basement of sedimentary rocks of Llandovery Age is exposed at the point and is unconformably overlain by breccia of Old Red Sandstone Age.

Dumfries and Galloway Region: Kirkudbright

Cairnsmore of Fleet (76)
Nr Newton Stewart
Kirkudbright
NX 500660 (OS sheet no 83)

Cairnsmore of Fleet is a mountain situated eight miles to the east of Newton Stewart. The mountain forms the south west end of a large granite intrusion of Early Devonian Age. This igneous mass consists of two types of granite, one bordering the other. The inner granite, which forms most of the intrusion, is a muscovite biotite granite while surrounding it is a mass of biotite granite. The intrusion has a large metamorphic aureole in Ordovician and Silurian sediments.

Access to the mountain is from the A75 Newton Stewart – Creetown road by way of unclassified roads from Blackcraig. This s inhospitable mountainous country and he normal safety precautions should apply.

Dumfries and Galloway Region: Lanarkshire

Wanlockhead (77)
Lanarkshire
NS 873129 (OS sheet no 71)

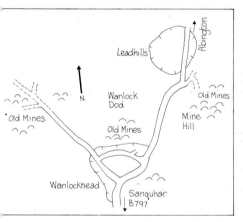

Wanlockhead, together with the neighbouring village of Leadhills, formed the centre of one of Scotland's most productive mining areas.

The geology of the area is described in two guides published by the Wanlockhead Museum Trust. The first guide entitled 'Wanlockhead's Rocks, Minerals and Fossils', describe the geology and mineralogy of the area, while the second guide entitled 'Wanlockhead's Mining History Trails' outlines informative walks in the area.

Old mine dumps are plentiful here and mineral specimens abound. Fifty-seven different minerals have been recorded including a number of rare varieties. Mining relics, together with rock and mineral specimens, can be seen in the Lead Mining Museum in the village, where the booklets can also be obtained.

Fife

Elie Ness (78)
Elie
Fife
NO 494000 (OS sheet no 59)

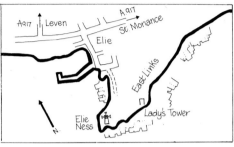

No fewer than five volcanic necks of Carboniferous Age are exposed on the shore around the small town of Elie.

Elie Ness is reached by following the harbour towards the east and turning right along a track towards the lighthouse. Ample parking space is available. The eastern margin of the Elie Ness volcanic neck and the baked country rock can be examined on the shore close to the Lady's Tower. Large areas of the neck are composed of bedded ash which is cut by dykes and calcite veins. Impure pyrope garnets are haphazardly distributed throughout the neck.

Lothian

Arthur's Seat (79)
Edinburgh
Lothian
NT 276730 (OS sheet no 66)

Arthur's Seat is the famous tourist attraction which lies within Holyrood Park. The hill is the remains of a volcano which was active during the Carboniferous Period. Volcanic vents, many of which are filled with agglomerate, can be seen and examined at Pulpit Rock, Crag's Vent, Castle Rock, Lion's Head and Lion's Haunch.

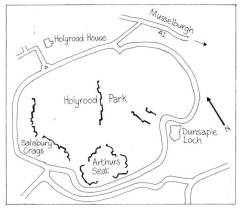

Salisbury Crags, a basalt sill, and the contact zone between this intrusion and the sedimentary rocks are also exposed.

North Berwick Law (80)
North Berwick
Lothian
NT 556843 (OS sheet no 66)

North Berwick Law is the eroded plug of a long extinct volcano. It is composed of trachyte, which has withstood the effects of erosion more easily than the softer country rocks surrounding it. In Scotland the name

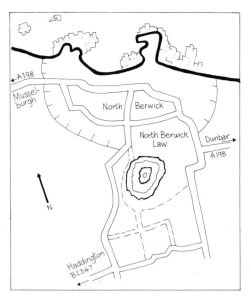

'Law' when applied to a hill often signifies the presence of a volcanic plug.

North Berwick Law stands 187m high and is situated immediately to the south of North Berwick between the B1347 and the A198.

Roxburghshire

Chiefswood (81)
Eildon Hills
Nr Melrose
Roxburghshire
NT 535336 (OS sheet no 73)

The Chiefswood Volcanic Neck is exposed in an old quarry in the Eildon Hills, south of Melrose. The quarry is reached by taking the road past the cemetary on the outskirts of the town.

This is a trachyte plug although basalt, agglomerate, sandstone and shale are also exposed in the quarry. The plug is of Carboniferous Age.

Eildon Hills (82)
Nr Melrose
Roxburghshire
NT 550325 (OS sheet no 73)

The Eildon Hills form the impressive range immediately to the south of Melrose. The hills are the eroded remnants of an intrustion of trachyte. It is considered that the magma was intruded into Late Devonian rocks as part of a volcano in Carboniferous times. The Eildon Hills are the centre of a complex of igneous rocks which take the forms of dykes, sills and vents.

Strathclyde Region: Ayrshire

Ailsa Craig (83)
NS 020000 (OS sheet no 76)

Ailsa Craig is a plug of Carboniferous Age and is made of a fine grained microgranite. On the south and west the granite is cut by numerous dolerite dykes which have weathered more rapidly giving rise to a number of caves. Courses on the geology of Ailsa Craig are organised by the National Trust for Scotland from Culzean Country Park near Maybole, Ayrshire.

Culzean (84)
Ayrshire
NS 230102 (OS sheet no 70)

Culzean Castle, which is the property of the National Trust for Scotland, stands upon basaltic and andesitic lavas of Devonian Age. Dykes of Devonian and Tertiary Ages are well exposed in the castle grounds.

The geology of the area is described in the guide entitled *'Culzean, Geology and Scenery*, which is available from Culzean Country Park, near Maybole, Ayrshire. The guide is published by The National Trust for Scotland.

Cumnock (85)
Ayrshire
NS 580205 (OS sheet no 70)

Carboniferous Limestone is exposed in the Shield Old Quarries which are situated approximately three miles to the south of Cumnock. This is another fossil locality, possible finds including corals, brachiopods and lamellibranchs.

Heads of Ayr (86)
Ayrshire
NS 284188 (OS sheet no 70)

The Heads of Ayr dominate the coastal scenery south of the City of Ayr. They are reached at low tide by walking southwards along the beach from Ayr, or northwards from the village of Dunure.

The Heads are the remains of a volcanic vent which was active during the Carboniferous Period. The vent, which stands 61m high and is almost a kilometre in length, is composed of agglomerate, but tuffs and lavas are exposed locally. Pebbles of the semi-precious stones agate and chalcedony may be found on adjacent beaches.

The Knipe (87)
New Cumnock
Ayrshire
NS 658103 (OS sheet no 71)

Antimony ores are scarce in Britain and have rarely been worked commercially, but one such location is situated near the summit of 'The Knipe'. The path which leads up to the old mine workings is long, hard and frequently boggy. Spoil heaps associated with the mine yield specimens of stibnite, usually with quartz. The stibnite, which occurs as grey radiating needles, has been worked from within a small granite mass.

Mauchline (88)
Ayrshire
NS 500272 (OS sheet no 70)

Permian Sandstones are well exposed in the Ballochmoyle Quarry. These sandstones were laid down during desert conditions and exhibit dune bedding.

Troon (89)
Ayrshire
NS 345343 (OS sheet no 70)

The Hillhouse Quarry was opened in an igneous sill. This sill is composed of alkali basalt and is well exposed within the quarry.

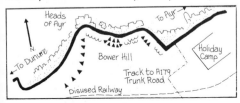

Woodland Point (90)
Girvan
Ayrshire
NX 170953 (OS sheet no 76)

The Ordovician and Silurian sedimentary rocks exposed here and in old quarries near Ardwell Farm, contain a variety of fossils. These include graptolites, brachiopods and trilobites.

Tayside Region: Angus

Auchmithie (91)
Angus
NO 680443 (OS sheet no 54)

Auchmithie is an attractive village on the coast of Angus, five miles to the north of Forfar. The village stands upon tall cliffs

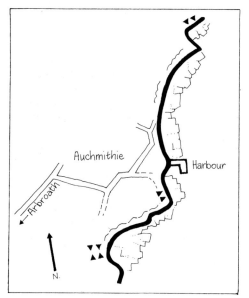

of conglomerate of Old Red Sandstone Age. The conglomerate contributes many pebbles to the beach here including large quantities of red jasper – Britain's most abundant semi-precious stone.

Fishtown of Usan (92)
Nr Montrose
Angus
NO 726546 (OS sheet no 54)

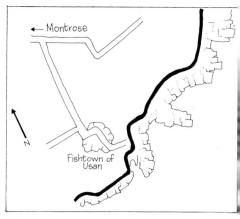

The cliffs at Usan are basaltic lavas of lower Old Red Sandstone Age. The lavas are well exposed and are frequently amygdaloidal, the vapour cavities having been filled with chlorite, calcite, quartz, chalcedony, agate and onyx. On occasion, the quartz is purple amethyst, or brown smoky quartz, both in crystal-lined geodes.

A channel formed in the rock platform below the village is where a dyke has weathered preferentially to the surrounding lavas.

Lunan Bay (93)
Angus
NO 690516 (OS sheet no 54)

The beach at Lunan Bay offers an interesting array of shingle which includes a variety of semi-precious stones. Agate, chalcedony and quartz pebbles occur, having been eroded out of the relatively soft amygdaloidal lavas around Montrose.

Pebbles from the metamorphic rocks of the Scottish Highlands also occur on this beach, transported to the area by glaciers. Garnets have been found in pebbles of mica schist.

NORTH SCOTLAND

Geology

The area comprises the mountains of the Highlands and the older continental rocks of the north-west foreland and the Outer Isles. The Highlands were, 400 million years ago, a massive mountain chain, probably with peaks of six or seven thousand metres. These were eroded down and only recently – in the last fifty million years, re-uplifted. We now see, at the surface, rocks that were once buried down below the Caledonian mountains – metamorphic rocks – schists and gneisses made out of the thick piles of pre-Cambrian Dalradian and Moine sediments. These were intruded by many granite and gabbro bodies and contain the remains of some overlying Devonian volcanoes – as around Ben Nevis. The whole region contains highly complex structures formed during mountain making. The big split down the centre of the area is the Great Glen, that follows the softer rocks of a major fault zone. Beyond the Moine thrust, in the North West, there are the Lewisian gneisses – rocks that have suffered many stages of uplift and erosion since they were formed more than 2,800 million years ago. Above this remnant piece of old continent there lies a considerable thickness of Torridonian (pre-Cambrian) sandstones and in places, Cambrian limestones that give a surprising 'Englishness' to the scenery of this remote part of Scotland. In the Tertiary, large thicknesses of lava were extruded on to some of the Inner Hebrides as well as parts of Northern Ireland and the remains of the accompanying giant volcanoes can be seen at a number of the islands and on the Ardnamurchan peninsula. Dykes cross-cut much of the area.

In the east of Scotland, there are thick red sandstones, conglomerates and shales of Devonian Age piled on top of the eroded Caledonian mountains. These fringe the Moray Firth and make up most of the Orkneys and some of the Shetlands.

Minerals

The metamorphic rocks of the Scottish Highlands hold much promise but little of practical value for the collector.

Garnets are very common in the schists of the Highlands, especially around Balmoral in Aberdeenshire and Bettyhill in Sutherland, but these are rarely larger than a pea in size. Crystals of the blue mineral kyanite also occur but are usually of the most inferior quality.

Volcanic rocks of Devonian Age form the hinterland around Oban on the west coast of Scotland, and agates occur in these rocks on Ben Nevis near Fort William. The Tertiary lavas also contain semi-precious stones; on Skye, near Dunvegan, and on the south coast of Mull, agates may be found as amygdales within the lavas. On Rhum, at Bloodstone Hill, agates occur with bloodstone.

Zeolite minerals abound in the volcanic rocks. Good places to search are near Storr and at Talisker Bay on Skye, and on the banks of Loch Scridain on the Island of Mull.

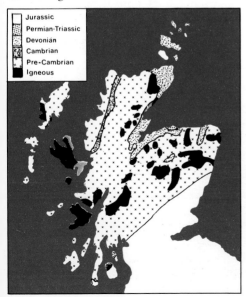

Jurassic
Permian-Triassic
Devonian
Cambrian
Pre-Cambrian
Igneous

Good stones for polishing, serpentine and marble, occur on several of the islands; those of Iona and Skye being the most famous. Another mineral occurrence of particular interest is the 'sapphires' of the Island of Mull. These may be found by the Carsaig Arches and near Tiroran on the north bank of Loch Scridain. Unfortunately the sapphires are impure and not suitable for polishing.

Shetland has a rather more interesting mineralogy than Orkney and is composed of sedimentary, igneous and metamorphic rocks. The most interesting locality is at Hillswick, where fluorite, kyanite and the unusual green garnet uvarovite occur.

Grampian Region: Banffshire

Cairngorm Mountains (94)
Banffshire
OS sheet no 36

The granite scenery of the Cairngorm Mountains makes up one of the most popular areas with tourists. Access to the mountains is most easily achieved from Aviemore, and a chair lift is available to convey passengers up the lower slopes of Cairngorm.

Pegmatites are known to occur within the granite and large crystals of the gemstone smoky quartz have been located here in the past. Although even small specimens are scarce today, the occasional weathered crystal may be encountered on screes in the granite area.

A small booklet is available entitled 'Cairngorm's National Nature Reserve', which includes brief details on the geology and natural history of the mountains. This booklet may be obtained from The Nature Conservancy, Achantoul, Aviemore, Inverness-shire.

Portsoy (95)
Banffshire
NJ 590662 (OS sheet no 29)

The small town of Portsoy on the Moray Firth is the home of the Portsoy Marble, actually a *serpentine* rock, that outcrops as cliffs immediately to the west of the town and occurs as pebbles on the beach.

Brown, red and green in colour, this attractive rock was used to decorate the palace of Versailles in France. The serpentine was formed by the metamorphism of a peridotite (olivine rich) igneous rock. Talc, which is found as an alteration product of serpentine, occurs in crush zones of the Portsoy serpentine and has been worked commercially.

Garnets occur in mica schist on The Battery, a small headland to the east of Portsoy.

G17. The Cairngorm mountains

Tynet Burn (96)
Fochabers
Banffshire
NJ 340590 (OS sheet no 28)

Sediments of Middle Old Red Sandstone
(Devonian) Age are exposed by the bend
in the burn, below the sawmills. The rocks
exposed belong to the Tynet Fish bed and
yield a variety of fossils including fish
remains.

Highlands Region: Argyll

Glen Coe (97)
Argyll
NN 290550 (OS sheet no 41)

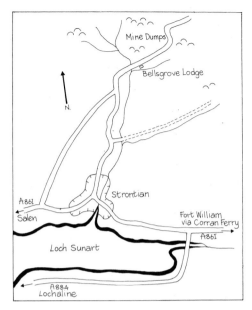

This is a very beautiful area of Scotland
and is particularly interesting geologically.
Glen Coe was a valley formed during
Tertiary times that became modified by the
glaciers during the recent Ice Age.

Glen Coe is of special interest through
containing *Cauldron Subsidence* that formed
in Devonian times due to the collapse of a
volcano and the formation of a caldera. A
granitic intrusion has formed around the
sides of the collapsed block. The heart of
Glen Coe, including the famous Three
Sisters, is hewn out of volcanic rocks of
Devonian (Old Red Sandstone) Age,
which have been protected from erosion
by the subsidence.

The relationship between the volcanic
rocks and the granite can be examined on
the flank of Ant-Sron.

Strontian (98)
Argyll
NM 830659 (OS sheet no 40)

The village of Strontian on Loch Sunart
was the centre of a large mining industry
during the 18th and 19th Centuries. The
mineral veins which contain the sulphide
ores of lead and zinc occur within a large
granodiorite intrusion. The mineral
strontianite occurs here and was named
after the town. It is the mineral from
which the element strontium was first
derived.

Spoil heaps associated with the old mine
workings are plentiful and can be found
immediately to the north of Bellsgrove
Lodge on the Strontian to Polloch road.

Highlands Region: Caithness

Achanarras (99)
Spital
Caithness
ND 150545 (OS sheet no 12)

The small, and now disused, quarry at
Achanarras is famous for the exposures of
highly fossiliferous limestones.

Achanarras Quarry is situated north of
Ballone and is reached by a rough track
which runs from the B870 Westerdale to
Mybster road. Permission can be obtained
to enter the quarry from Mr J Saxon,
c/o Thurso Library, Thurso, Caithness.

The impure limestones exposed in the
quarry are of Middle Old Red Sandstone

(Devonian) Age and contain numerous examples of fossil fishes. The minerals galena, sphalerite, calcite, quartz and pyrite may be found within the quarry.

Duncansby Head (100)
Nr John O'Groats
Caithness
ND 406734 (OS sheet no 7)

From the car park at Duncansby Head there is an attractive walk along the Cliffs to the south.

Duncansby Head, which stands 60m high, is formed from the Thurso Flags, while to the south the John O'Groats Sandstone makes up the cliffs. These rocks, of Middle Old Red Sandstone (Devonian) Age, are well exposed and the Sandstone forms the magnificent Duncansby Stacks which dominate the cliff walk. Such *sea stacks* form by wave erosion when two caves on the opposite sides of a headland join up and the roof, or arch, falls in.

Highlands Region: Inverness-shire

Island of Eigg (101)
OS sheet no 39

The geology of the Islands of Eigg and Muck is described in a guide to the islands by L MacEwan. This guide can be obtained from the Tourist Information Office, Mallaig, Inverness-shire.

Island of Rhum (102)
OS sheet no 39

The Island of Rhum was one of the major volcanic centres during the Tertiary volcanic episode. Now the property of the Nature Conservancy, the geology of the island is described in a conservancy publication.

Day visitors to the island from Mallaig may enjoy the two nature trails provided.

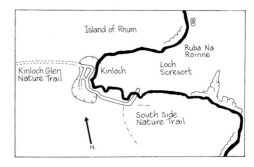

These are the Kinloch Glen and the South Side Nature Trails. A guide to the nature trails, together with the geological guides, can be obtained from The Chief Warden, The White House, Island of Rhum.

Island of Skye (103)
OS sheet nos 23 and 32

The Tertiary Igneous Geology of the Island of Skye is described in the Geologists' Association Guide No 13.

Mallaig (104)
Inverness-shire
NM 676970 (OS sheet no 40)

A guide to the Lewisian Gneisses and Moine Schists around Mallaig is published by the Geologists' Association in Guide No 35 by R St J Lambert and A B Poole.

G19. The Skye Marble quarry near Elgol on the Island of Skye

Knockan Cliff (105)
Inverpolly
Knockan
Ross-shire
NC 187087 (OS sheet no 15)

This is one of the most notable localities of geological interest in Britain. The cliff contains not only exposures of Britain's oldest rocks but also reveals the highly altered Moine Schist overlying the unaltered Durness Limestone. This irrefutable evidence of the Moine Thrust is described in detail in the guide to the *Knockan Cliff Geological Trail* published by the Nature Conservancy Council obtainable from The Warden, Strathpolly, Inverpolly by Ullapool, Ross and Cromarty.

Loch Maree (106)
Kinlochewe
Ross-shire
NH 001650 (OS sheet no 19)

The geology of the area around Loch Maree is explained in the guide to 'The Glas Leitire Nature Trail' which begins from the Loch Maree Picnic Site three miles from Kinlochewe on the Gairloch road. The guide is published by the Nature Conservancy, 12 Hope Terrace, Edinburgh EH9 2AC.

318. *Tertiary lavas of the Island of Skye. Massive landslipped fragment of these, centre left*

Highlands Region: Sutherland

Inverpolly (107)
Sutherland
NC 110118 (OS sheet no 15)

Details of geological interest are described in a Nature Conservancy guide to the Inverpolly area. The guide which describes the Inverpolly Motor Trail is available from the Nature Conservancy, 12 Hope Terrace, Edinburgh EH9 2AS.

Kildonan (108)
Nr Helmsdale
Sutherland
NC 912210 (OS sheet no 17)

Kildonan was the scene of a gold rush in 1861 and gold still occurs amongst the river gravel in the Kildonan and Suisgill Burns. The gold is alluvial and with a little practice can be panned quite easily. Permits to pan for gold must be obtained from the Suisgill Estate.

Lochinver (109)
Sutherland
NC 107140 (OS sheet no 15)

The oldest rocks in the British Isles, the Lewisian Gneiss, are well exposed on the hillside adjacent to the Loch na Muirichinn. Garnets, usually very small, occur in bands within the gneiss. Reach the rock exposures on the Lochinver–Inverkirkaig road.

North East Scotland (110)

The Dalradian rocks of North East Scotland are described in the Geologists' Association Guide No 31 by H H Read.

Strath Naver (111)
Betty Hill
Sutherland
NC 706615 (OS sheet no 10)

The Moine schists are well exposed on the west side of Strath Naver near the mouth of the river Naver. Small grains of garnet occur within bands of the schist at this location.

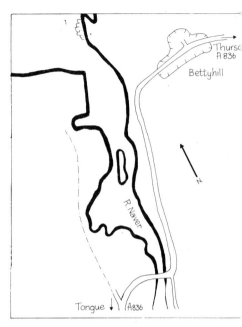

Strathclyde Region

Carsaig Arches (112)
Island of Mull
Argyll
NM 500186 (OS sheet no 48)

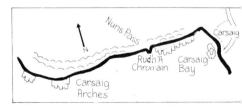

The cliffs on the south coast of Mull are predominantly composed of basaltic lava of Tertiary Age. The lavas are frequently amygdaloidal, the amygdales being composed of either quartz, chalcedony, agate, or one of the zeolite minerals. 'Sapphire' bearing xenoliths have been located in a number of sills which have enetrated the lavas near Rudh 'a' Chromain, west of Carsaig. Although of considerable mineralogical interest, the sapphires are impure and badly fractured, and consequently have no decorative or commercial application.

Corrie (113)
Isle of Arran
NS 024433 (OS sheet no 69)

At Corrie the shore section is an exposure of Carboniferous limestone and sandstones which are highly fossiliferous. The main fossils are of plant and fish remains and corals.

320. Torridonian sandstones resting on the Lewisian – Sliogh and Loch Maree

Goat Fell (114)
Isle of Arran
NR 990415 (OS sheet no 69)

The mountain of Goat Fell is carved out of a coarse grained granite intrusion of Tertiary Age. Druses (cavities) lined with crystals are not uncommon and on occasions crystals of amethyst, smoky quartz, citrine, rock crystal, blue beryl and topaz may be found.

The contact zone between the granite and the surrounding Dalradian Schists can be examined at several localities in Glen Rose and in North Glen Sannox.

An intrusion of fine grained granite is exposed within the coarse grained granite around Loch Tanna, the contact zone between the granites being visible in Glen Catacol.

G21. Columnar jointing in the Tertiary basalts – Fingal's Cave on the Island of Staffa

Gribun (115)
Island of Mull
Argyll
NM 450335 (OS sheet no 48)

Rocks of Cretaceous Age, corresponding to the Greensand Series, are exposed approximately 320m to the north east of Balmeanach Farm, below the road. This is an unusual exposure amongst the Tertiary volcanic rocks of Mull, the rocks yielding various examples of brachiopods and lamellibranchs.

Island of Mull (116)
OS sheet nos 48 and 49

The Tertiary Igneous geology of the Island of Mull is described in the Geologists' Association Guide No 20 by R R Skelhorn, J D S MacDougall and P J N Longland.

Island of Staffa (117)
Inverness-shire
OS sheet no 47

The columnar basalt which surrounds Fingal's Cave, on the Island of Staffa, is one of the best examples in the British Isles. The island can be visited by motor launch during the summer months, from Ulva ferry on the Island of Mull.

Isle of Arran (118)
OS sheet no 69

The geology of Arran is described in the Geologists' Association Guide No 32 by S I Tomkeieff.

Lochranza (119)
Laggan
Isle of Arran
NR 975500 (OS sheet no 69)

Carboniferous rocks which include Calciferous sandstone and limestone are exposed to the north and south of Laggan. The exposures yield many examples of lamellibranchs, trilobites, corals and fish remains.

Appendix

Mineral Collecting in Britain

During recent years, collecting minerals and gemstones has become one of the most popular hobbies in Britain. Beaches are no longer deserted in the winter time, and old quarries and mine dumps, abandoned by man many decades ago, are once again centres of human interest.

Collecting in one form or another is so widespread, in fact, that it begins to look like a fundamental aspect of human nature. The objects of the collecting 'instinct' may differ but collecting, whether for pleasure or for profit, can be a stimulating and exciting pastime.

What the collector of minerals dreams of is discovering the 'perfect' specimen, but perfection in minerals is, as in most things, hard to find. Even without it, however, minerals are the products of the earth and in many instances were created millions of years ago, so they are antiques on a very grand scale. We may experience the pleasure of collecting them but in the end we are custodians, nothing more.

The geology of the British Isles is very complex and minerals of many varieties occur somewhere within our shores. Unfortunately for the collector they very rarely achieve the beauty and splendour of specimens from other lands, largely because the mines in Britain were exploited so long ago that the best material has gone. True, rare and unusual minerals may be found in Britain, but they may be represented by a crystal little more than one millimetre in length, or by a grain so small it can only be seen when magnified by a microscope. So, as most collectors seek to collect large and well formed examples of Britain's minerals, their search must inevitably be for the more common varieties.

Britain also has its share of stones which are worth collecting simply because they are aesthetically pleasing. The name gemstone may apply to any rock or mineral which is attractive and can be utilised for decorative purposes. Gemstones usually require polishing before they exhibit their decorative qualities and not surprisingly many collectors are also experts at stone cutting and polishing, which is the craft called lapidary.

Some gemstones are more common than others. Serpentine, for example, is a rock which occurs in large quantities in Britain; carnelian on the other hand is scarce, and in many instances may only be found as small pebbles on a beach. The scarcer gemstones have earned the name 'semi-precious stones' but in all honesty, with the possible exception of agate, few specimens of true semi-precious quality can be found in Britain.

Collecting and Conservation

Collecting minerals and gemstones is, on the face of it, a harmless enough pastime but there are problems. For instance, one person taking a mineral specimen from a location will do little harm, but if a thousand people each take one specimen from the location, little of interest may remain. In short, collectors have a responsibility to society and to themselves, to preserve geological locations.

The collector can play a role in conservation if by collecting a specimen he prevents it from being destroyed. Beaches, screes on hillsides and old mine dumps are places where specimens of minerals and gemstones are being destroyed due to erosion by the elements. These erosive forces can quickly disfigure and subsequently destroy the softer minerals and even hard minerals eventually suffer the same fate. Collecting minerals at localities such as these is doing little harm.

Collecting by hammering on rock surfaces is to be discouraged for several reasons. Damaging a rock exposure scars the landscape, helps increase the erosion of the exposure, and ultimately destroys the locality, together with any scientific value it might have held. In any case it is far more sensible to let Nature do the hard work and limit ourselves to picking up what she has been kind enough to provide.

Code for Amateur Geologists and Mineral Collectors

The following code is designed to encourage the best possible relationship with site owners, preserve localities, and outline areas of potential danger at geological sites.

1 Always obtain permission to enter private locations in advance, and adhere to any instructions given by the owner.
2 Always follow the country code. Neglecting to close gates, trampling crops, and taking dogs into fields containing sheep, are actions which bring the geological fraternity into disrepute. Relations with site owners can be irreparably damaged by such actions.
3 Leave the locality in a safe condition such that other visitors can explore the site in complete safety. Do not leave behind hazards, even tins and plastic bags, which could injure farm animals.
4 In quarries or gravel pits avoid walking near the working face or on recently blasted rock which may be unstable.
5 When exploring old mine dumps beware of unguarded shafts. Many of these shafts which are hundreds of feet deep are still inadequately covered.
6 Consult tide tables prior to exploring beaches, especially those backed by tall cliffs. Avoid walking under the cliffs.
7 Do not hammer on exposed rock faces.
8 Only collect minerals and gemstones from locations such as beaches, mine dumps or working quarries where specimens would otherwise be destroyed by nature or man's commercial activities.
9 Never collect more material than you really need. Leave some for the next person.
10 If you find a specimen which has unusual characteristics please take it to a museum or other institution where it can be positively identified, and any unusual features noted.

Forming a mineral collection

Without the storage space of a museum it is hardly practical to collect examples of all Britain's rocks and minerals, so it is necessary for most collectors to specialize in some particular aspect of mineralogy. You may choose to specialize in gemstones or ores of metals, or you may adopt a more technical theme such as 'carbonates' or 'silicates' and try to collect examples of as many as possible. If the collection is to have any scientific value at all it is essential that the collector keeps an accurate record of each specimen in his collection. The memory cannot be trusted so it is necessary to note the following.

1 the date the specimen was found,
2 the locality with Ordnance Survey Grid Reference Number,
3 accurate name of specimen.

Help with identifying unusual specimens can usually be obtained at museums and universities.

Unfortunately many collections never see the light of day and this applies to museums and university collections as well as those in private hands. I personally believe that every opportunity should be taken to allow members of the general public the opportunity to appreciate the scientific and decorative value of any specimens you may have obtained. To this end I commend membership of one of the many amateur geological societies or lapidary societies which occur throughout the country. Not only can you share your interest with like-minded people, but also use the society as a platform from which to present knowledge to others who might become interested.

Maps of mineral localities appear on pages 134 and 158.   PRE

Index
of Geological terms

*Front cover: eroded
basalt lavas of
Storr, Isle of Skye*